Praise for *The Science of Paddling*

Shawn Burke uses his experience as an engineer and passion for canoeing to answer the questions many of us have always had about paddled hulls. He establishes a common language, based in science, with terminology, definitions, and explanations that provides all paddlers the means to explore and understand the complexities of paddlesport, and become more efficient so as to travel further, faster, and with less effort. Whether you are a weekend recreational paddler or a national level competitor, this book is for you.

> — HARRY ROCK, member of the ACA Paddlesport Hall of Fame and author of *The Basic Essentials of Canoe Poling*

The perfect book for the paddler with an inquiring mind and a passion for math and science. The questions and solutions explored in each chapter enlighten and empower the reader with knowledge to be applied next time on the water.

> — CHRIS PRATER, owner, Ripple FX Paddles

For paddling nerds like me, the articles that Shawn has written, initially on our website, and now in this compendium, have been fascinating. Want to know when it's important for your partner in a double to keep perfect time? Need confirmation that it's best that your partner goes on a diet – to fix the trim in the boat?! Want to know why a paddle works the way it does? It's all here – for competitive, serious recreational or simply curious paddlers, this book is full of truly interesting answers to the technical "whys and wherefores" of paddling.

> — ROB MOUSLEY, www.surfski.info

Here it is! The facts and players needed to build an understanding of the physics of moving paddlecraft. This book is clear, factual, and full of detail. The how, the why, and stories to educate our curious minds about the sport we love.

> — BOB BLAIR, Advanced Ship Handling Chief Instructor, Massachusetts Maritime Academy and owner, Speedboard USA

If you enjoy getting out on the water and paddling, be it in a canoe, kayak, SUP, surf ski, dragonboat, or outrigger, then *The Science of Paddling* is an absolute must read! Shawn Burke not only makes science understandable and interesting, but he explains how these concepts can be applied to help make any paddler better. This book is packed full with practical paddling tips and advice, allowing anyone to become more efficient on the water. You will learn not only how and why certain scientific principles apply to all canoes and kayaks, but most importantly, how you can use this knowledge to move any boat faster with less effort. Whether you are a newcomer to paddlesports or an experienced racer, *The Science of Paddling* is for you. See you on the water!

> — PETER HEED, co-author of *Canoe Racing: The Competitor's Guide to Marathon and Downriver Canoe Racing* and multi-time National marathon and downriver canoe champion

Acknowledgements

I am indebted to my content editors. Kirk Olsen and Todd Johnson are experienced paddlers and engineers who provided excellent feedback and on the water perspective to my writing. Dr. Charles "Chuck" Gedney, a research and consulting engineer and one of my MIT grad school office mates, ensured I got the fluid mechanics and mathematics right. All three encouraged me while offering improvements for this first edition.

I am especially indebted to my editor, Doug Berg. It should come as no surprise that writing a book is a slog. Doug waded through the sea of content and updates, keeping me focused on the reader at all times. I've been fortunate to race Doug a few times, and he brought the true competitor's spirit of "seeking together" to this project.

I am grateful for the reviewers who offered their comments and feedback for articles appearing on the Science of Paddling blog. Many of those articles appear in revised and expanded form here. Their perspective kept my writing grounded. Readers of the blog have offered comments and asked insightful questions, pointing out topics worth exploring. This led to some interesting posts and helped connect me with the worldwide paddling community.

Many readers and fellow paddlers encouraged me to write this book, opening new territory in paddlesport's *terra incognita* for us all to explore. Special thanks go to Benjamin Meader, who told me that there was a book to be written from my scattered musings. Thanks for providing the perspective I was missing and pushing me in the right direction.

I want to offer my gratitude to three members of the Contoocook Tuesday night paddling group. Phil Cole was a physicist who asked great questions that led to some of my earliest articles. He regaled me with endless tales of engineering and other adventures. Peter Beckett provided a boat builder and designer's perspective as we discussed my take on drag and propulsion.

And thanks to my dear friend, mentor, and paddling partner Tom Walton, who asked the question that started it all: "Why isn't a tandem twice as fast as a solo?" Thanks, Dude.

Disclaimer

We are not lawyers. This book and the content provided herein are for educational purposes only. Every effort has been made to ensure that the content provided is accurate and helpful for our readers at the time of publication. However, this is not an exhaustive treatment of the subjects covered. Neither the author nor the publisher claims responsibility for adverse effects resulting from the use of the information found within this book. No liability is assumed for losses or damages due to the information provided. You are solely responsible for your own choices, actions, and results.

We are also not medical doctors, and we are not offering medical advice. Be sure to consult with your physician before undertaking any exercise program, and before performing any of the performance tests or assessments outlined in this book.

For Monica

"There is a light that shines in my heart."

Contents

Introduction

One Summer evening in 2008, our Tuesday training group was paddling back to the boat ramp on the Contoocook River after a long session on the water. This night saw us moving together in a pack at a moderate pace, interspersed with sprints at turns in the river. We were idly chatting during a lull in the action when my friend Tom pulled alongside, turned, and asked me, "Why can't two people paddle a tandem canoe twice as fast as a solo paddled by one?"

I had no idea. But I knew that I should, and that bugged me.

As an engineer, I find paddling not only fun but technically fascinating. Spending hours at a time on an indoor paddling trainer in Winter and on the water when it's not frozen here in New England has given me time to ponder my sport. Tom's query resonated because I'm curious about why things work. Despite my education and training, I'd always glossed over the specifics for paddling until he posed that question. Canoes move when paddled; isn't that enough? Obviously, "science is happening," but how? I'd offered the occasional nod to Isaac Newton's 3rd Law of action-reaction, which requires that the stern-directed paddle force results in the hull moving forward. Or I cited Archimedes' "Eureka!" moment, revealing why particular combinations of weight and displaced hull volume predict some things will float while others sink. Beyond that, my understanding of the *why* was shallow. I knew more about what to do than why things happen.

From experience, I knew that an empty canoe (usually) passes unscathed through a series of rapids. A kayak slows down in shoals yet naturally speeds up as it passes into deeper water. Plus so many other phenomena that I had experienced, but mainly as a passive observer. The overarching questions became, *why do paddled hulls do what they do?* And is paddling fundamentally different than other sports in how it engages the body? The hull and the water know what to do – that's their nature. To understand what they know, to get at the *why*, we must learn their language: the language of science. My investigations cast a new light on years of paddling. What I learned was enriching and sometimes counter intuitive; that's where the fun is. And that's why I want to share what I've learned with you.

The Science of Paddling addresses the why question of paddled hulls. The concepts presented here apply to canoes, kayaks, surfskis, outriggers, and SUPs. All these craft have four things in common: a hull, at least one paddle, one or more paddlers, and water. I've limited myself to topics that are broadly applicable, rather targeting a specific hull model or brand of paddle. Consequently, what you'll learn applies to you and your craft.

Along the way you'll find ways to make your paddling more efficient, allowing you to travel further, faster, and with less effort. Competitive paddlers will see that physics can guide their training to be more paddling-specific. And finally, I hope that like me you discover the simple joy of understanding our favorite sport, and gaining a new appreciation for the wonder of gliding across the water.

Organization and Audience

In developing this book, I found that topics follow a progression, naturally divided into two portions. In the first part, Chapters 1 through 7, we construct the foundation for later applications, with topics building one upon the other. We address why a hull is buoyant, why it moves, what opposes its motion, and why a paddle propels it. Core concepts from classical physics are introduced and applied in context: Newton's Laws of Motion, energy conservation, and mass conservation. You'll see that these principles are merely precise statements that represent our everyday experiences of the world around us.

The foundation laid in the first seven chapters lets us then explore several applications. Topics include trim's effect on handling, how to measure your hull's drag coefficient, why we always lose time on an out-and-back route in current, how physics can ensure all paddlers are treated equally in race scoring, and how endurance and strength training can be optimized for our sport. Woven throughout are reflections on how engineers approach problem solving.

Each chapter is organized as follows:

- THE TOPIC: problem statement, modeling and analysis, and conclusions. Primarily text exposition, with analysis limited to algebra and geometry at most.
- TAKE-AWAYS: bullet points that summarize the main points from the chapter. Here, I offer pub quiz-ready talking points for the main questions addressed.
- EXTRA CREDIT: detailed modeling or analysis is presented for the interested reader – the general reader loses nothing by skipping over these parts or returning to them later.
- FURTHER READING: references are included for those who wish to dive deeper into the chapter's subject matter.

This structure lets readers explore the material at whatever depth they wish. You needn't have a degree in science, engineering, or mathematics to learn about the why of paddled hulls. Just an open and curious mind.

• • •

So here's what you'll find:

Chapter 0: Newton's Laws of Motion

This brief preface summarizes Newton's Laws of Motion. While many readers may have a working knowledge of classical physics, it's best to offer this review since the Laws underpin so much of what follows. Even experienced practitioners will benefit from seeing how Newton originally stated the 2nd Law – it perfectly synopsizes how paddles propel our craft.

Chapter 1: Tandem and Solo

Why aren't tandem canoes twice as fast solo canoes? Fundamentally, it's all because of fluid mechanical drag. Our canoes move through the water due to the force we apply through our paddles. The water exerts an opposing force against the hull, which slows us down. This drag force is proportional to the square of our paddling speed, something we can prove without solving the complex equations that describe fluid flow around a hull.

This insight lets us compare tandem and solo hulls in terms of their respective power plants: the paddlers. Our analysis shows that the speed of a paddled hull is nonlinearly proportional to the paddling power from the total number of paddlers in the hull. Consequently, a tandem will move less than twice as fast as a solo. We then refine our analysis to account for the different carrying weights of these hulls.

Chapter 2: Kind of A Drag

The drag force that opposes a hull's motion is the sum of friction drag, form drag, and wave-making drag. Most people are unaware of the latter two elements and assume that only friction slows them down.

When water flows past a hull's solid surface, the molecules contacting the surface stick. This phenomenon is called the no-slip condition. As you move a small distance away from the hull, the water is no longer stuck, though it moves more slowly than your paddling speed. Move further away, and the water flows freely. This gradient in flow speed adjacent to the hull defines a boundary layer. Near the bow, the flow is smooth; the flow over the rest of the hull is turbulent. These effects are the source of friction drag.

Friction drag is proportional to the hull's wetted surface area – the part below the waterline. This suggests a simple experiment. Hold your arm out straight, palm flat and fingers pointed, with your hand parallel to the floor. Now briskly swing your arm side-to-side and notice the force exerted on your hand by the air. Repeat this experiment with your hand perpendicular to the floor. The surface area of your hand is the same in both orientations, so the friction drag must be the same. Yet somehow, the force experienced by your hand is greater when the airflow is perpendicular to the palm. What you just experienced is form drag. Form drag is the force that arises because fluid – in this case, air – flows past an object, and the object changes the flow's momentum. The change in momentum across your hull's wake is the source of form drag.

Chapter 3: Wave Drag and Shallow Water

Canoes, kayaks, and SUPs inhabit a unique environment: the interface between water and sky. This interface supports wave propagation. These waves are a little bit like the vibrations of a plucked guitar string. But in deep water, longer waves travel faster than shorter ones, and vice versa in shallow water. Our hulls must conform to these rules, which is where wave drag enters.

Hulls create waves because of energy conservation. Slower water near the bow has more potential energy; waves next to the hull grow tall there. Faster water further along the hull has more kinetic energy, creating a trough since potential energy there must be less than at the bow. All this energy comes from you, the paddler. The resulting wave pattern looks to synchronize with waves on the water's surface. The water naturally "tunes" itself to particular depth-dependent patterns of waves. And like the vibrations of a guitar string, water waves don't encourage playing off-key from their tuning.

Chapter 4: What Moves You

Why does a hull move when paddled?

One view is you exert a paddling force, and the hull moves in the opposite direction; we know this from experience. Another way of looking at it is each paddle stroke imparts motion to a mass of water behind and around the paddle blade. With each stroke, you "throw" this entrained water behind you. Since momentum is conserved, the momentum you and the hull achieve must equal the momentum imparted to the water. This determines your speed. We can use this model to show that you expend less energy paddling with a slightly larger blade than a smaller one. A larger blade, however, may necessitate a stronger pull.

Chapter 5: Roll Your Boat

Why do our hulls float in the first place? To answer that, we owe a debt of gratitude to Archimedes. He showed that a floating object displaces a mass of liquid that weighs the same as the object. For paddled canoes and kayaks, this corresponds to the portion of the hull below the waterline. And we know from Isaac Newton that floating equilibrium occurs because there is a balance of forces. The downward-directed force is the hull plus paddler(s) weight, and the upward restoring force equals the weight of displaced water. This restoring force is represented as a localized force directed upward from the displaced water volume's center of mass.

This center of mass moves as you lean the hull – the shape of the displaced water changes and moves toward the leaned side. The restoring force moves there as well. If this force lies outside of the downward force of your weight, the hull remains stable. But if you lean outside the restoring force, you soon will be swimming.

Chapter 6: About the Paddle

Paddles are essentially miniature barn doors that we sweep through the water to propel our craft. The paddling force they impart is akin to a drag force – here, mostly form drag owing to the blade's size and shape. The faster we pull it through the water during the stroke's power phase, the greater the force. And the paddling force depends on the square of the speed with which we pull the blade.

The paddle's propulsive efficiency depends on two factors. First, maximize the paddling force when the blade is perpendicular to the water's surface. And second, maximize the product of the average propulsive force and the stroke's duration. The geometric constraint leads us to explore so-called "bent shaft" paddles, where the paddle blade is set at an angle to the shaft to find an optimal bend angle.

Chapter 7: Impulse

What is the minimum number of measurable quantities for modeling paddling efficiency? Intuitively, we might think that if each stroke is longer and the number of strokes per minute (aka, cadence) is higher, we paddler faster – which is, of course, true. But a faster cadence and longer distance per stroke result from better underlying mechanics.

Over a single stroke, the product of average blade force and the stroke's duration equals impulse. Impulse quantifies how a single stroke increases the momentum of the hull and paddler(s) – remember, our hulls move because of momentum conservation. Blade impulse is the input. Propulsion occurs during the stroke's power phase. The stroke's duty cycle equals the fraction of time over the entire stroke where you're exerting a force. And the number of strokes over a given time is the cadence. The product of these three parameters equals the momentum imparted by the paddler(s). This model lets us optimize the propulsive force by varying these quantities.

INTERLUDE: Tales of Power

Before we move into the applications part of the book, we'll take a scenic turnoff.

We don't see enough science and engineering articles about why or when things don't work out. While writing *The Science of Paddling*, there have been a lot of hen scratchings, simulations, and data explorations that ended up on the proverbial cutting room floor. That's why it's fun to show how I fell down the rabbit hole searching for a way to measure a paddling stroke's power. In their landmark paper, "Sound waves in rooms," Philip Morse and Richard Bolt wrote of a need to "salt our analysis with liberal doses of common sense." Yet it isn't easy to let go of an assumption even when it violates fundamental physics. Or admit that a "final" design is only useful in highly controlled and unrealistic situations. It's an object lesson: to develop a meaningful (or even workable) engineering system, all relevant factors must be accounted for. You know, solid engineering practice 'n stuff. Plus salt.

Chapter 8: There and Back

Which is faster: paddling an out-and-back route without current or with the current? You lose time going upstream into current, but you regain it going downstream... don't you? A round trip should take the same amount of time, right?

Nope.

We tend to think of the round-trip transit problem in terms of distance since we remember that distance equals velocity times time. Or we think of it in terms of velocity. If we double the velocity we go twice as far for a fixed time interval. Simple! But for a fixed distance course, time and velocity have an inverse relationship. While you speed up on the downstream current, you recover less time going downstream than you lose fighting current going upstream. By shifting perspective to consider time, you see that you lose time paddling in current no matter what.

Chapter 9: Many Rivers to Cross

You're resting in slack water along the shore of a whitewater river, wondering how to work across the current to get into the next eddy. You're tired after a day of running rapids and wondering, "Gee, I wonder what the fastest way to do that is?" And if you're like me, you might ask, "Can I get there in the shortest distance, too?"

Well, sure, you can do that – you've posed two optimization problems. And you can solve them using pictures – simply by placing velocity vectors end-to-end. This avoids having to learn variational calculus; it's a win-win.

Chapter 10: Trim

Have you ever wondered why you struggle with control of your canoe when underway? Have you asked yourself why you should distribute weight in the hull to affect a particular trim? Have you felt your speed was lacking in certain water conditions? This chapter uses physics to understand the impact of changing your hull's underwater shape via trim and how trim can facilitate improved control, speed, and efficiency when paddling. To do this we utilize wake and wave drag concepts developed in Chapters 1 and 3.

Chapter 11: A Real Drag

I'd often wondered if the performance of a canoe, kayak, or SUP can be characterized by just a few numbers – like the information printed on a new car's window sticker. Physics provides a number that may be useful in characterizing and even comparing hulls.

The drag force can be summarized by a single parameter: the drag coefficient. It is a property of each unique hull shape and reflects how it sits in the water. Using GPS data from a series of field tests lets us compute a hull's drag coefficient. We consider drag data for a popular solo canoe and find that it embodies the square-law dependence of drag versus speed developed in Chapter 1.

Chapter 12: Start Me Up

We'll utilize the drag data from Chapter 11 to construct a simple simulation model of a solo canoe. Using MATLAB we compare various paddling propulsive force profiles to determine the "best" way to bring a hull up to speed. This chapter lets us close the loop and see how Newton's 2nd Law of Motion complements our understanding of momentum conservation and impulse from Chapters 4 and 7.

Chapter 13: Cutting Corners

When is it advantageous in a race to cut a shallow water corner? A primary difference between deep and shallow water is wave drag. Using an equal-effort metric, we find that the inner radius you can paddle to keep pace with a hull in deep water varies as the square root of depth. If the "shallow" water is deeper, the problem simplifies to picking the shortest path using geometry.

Chapter 14: Speed Above Replacement

Is there a way science can help us determine whether a paddler is equal to, lesser, or greater than the other paddlers in a boat, akin to baseball's "Wins Above Replacement"?

In Chapter 1, we learned that the average speed of a paddled hull is proportional to the cube root of the summed paddling power of all paddlers in the boat. If we consider a

4-person hull, then taking one paddler offline reduces the steady-state hull speed by 10%; for a 6-person outrigger, the speed reduction will be 6.2%. If the measured speed reduction is more than this, the subject paddler outperforms their peers; if less, they are underperforming. As the number of paddlers in the hull increases, this factor will decrease. Eventually, it falls below the measurement error of most GPS devices. We develop a test protocol to measure this including the effects of current and wind.

Chapter 15: Leveling the Field

One of the joys of paddlesport racing is seeing so many different people and boats at an event. You can compare their race results by adjusting their finishing times based on age, gender, and hull type. These time adjustments level the field so that their results can be compared and scored fairly. But what if there is not enough historical data upon which to base a time adjustment? We use the nonlinear dependence of average hull speed on power to fill in insufficient data, thus leveling the field for all paddlers.

Chapter 16: What Fuels You

When preparing for the racing season, why should I train in any particular way? The common wisdom is that long paddling workouts build endurance. We perform short, hard intervals to enhance our aerobic metabolism and tempo work to strengthen anaerobic capacity. But the fundamental question remains: Why?

This chapter explores the three metabolic pathways that fuel our working muscles: the aerobic, anaerobic, and phosphate systems. This helps us understand why specific training prescriptions work, and level sets us for the following three chapters.

Chapter 17: Power to the Paddlers

And as we learned in Chapter 1, paddling power determines the average cruising speed of a hull. Power is the application of a force quickly. Paddlers do this over and over (and over and over). So, if you intend to strength train for paddling, which do you focus on, force or power?

The branch of strength training called Velocity Based Training ("VBT") focuses not on how much iron you lift but how fast you move it. VBT posits velocity training ranges that provide specific functional benefits. These benefits accrue irrespective of your single-repetition maximum lifts since you target a velocity rather than a weight. We'll establish an optimal velocity for developing paddling power, then outline example paddling VBT training routines.

Chapter 18: The Deflection Point

Heart rate is a measure of exercise intensity: the harder you work, the faster your heart beats. For sports like running, the relationship between heart rate and intensity is linear up to the lactate threshold. Above that point, a plot of speed versus heart rate bends downward at the so-called deflection point. Runners can measure and plot their speed versus heart rate in a controlled track test to determine where their deflection point occurs.

Unlike running, the resistance paddlers experience versus speed is nonlinear. This means a speed versus heart rate plot won't follow a straight line; the nonlinear drag effect also muddies the deflection point location. Our understanding of paddling physics lets us remove the nonlinearity, making the deflection point easier to identify from test data for most paddlers.

Chapter 19: Paddling 30-30 Intervals

Exercising or racing at our maximum aerobic capacity builds that capacity and breaks our bodies down. Dr. Veronique Billat developed a protocol for improving runners' ability to process oxygen without undue stress on their bodies or prolonged recovery time. The protocol consists of short, sharp periods of exercise (aka, intervals) at maximum effort, interspersed with short periods of recovery where our metabolism continues to function at this maximum level. During recovery, the exercise is done at one-half the speed, or level of effort.

If paddlers employ Billat's interval workout, their level of effort during the recovery phases will be too low. Accounting for the cubic dependence of power on speed from Chapter 1 lets us derive a recovery speed for paddlers that matches her exercise prescription. We can then get the most out of our interval workouts.

Appendix: Reading Mathematics

As the late physicist Stephen Hawking noted in reference to his book *A Brief History of Time*, "Someone told me that each equation I included in the book would halve the sales." Yet his book sold an astounding 25 million copies from 1988 through 2007. I'm not Stephen Hawking. But it's hard to avoid a dose of mathematics when discussing the science of paddling since the material is grounded in physics, and the language of physics is mathematics.

"All that math" needn't be a mystery. Mathematics is just another language, albeit one that might not be very familiar. Like all languages, math has rules. Learning just enough of its grammar, words, and phrases will help you navigate the landscape, which is the goal of this Appendix. This primer facilitates a more immersive experience as you work through technical articles and books like this one. The goal isn't to do math but to read and get more from it. As noted chef and author Alton Brown often says, "Your patience will be rewarded."

Part I

Foundation

CHAPTER 0

Newton's Laws of Motion

Introduction

Quick: Name Newton's Laws of Motion. Are you able to recite all three? (Do you know there are three?) Are you comfortable describing at least one practical example of each?

These questions aren't intended to put you on the spot. But most of *The Science of Paddling* relies upon the Laws of Motion to explore the why of paddlesport. The good news is that the Laws are empirical. They were developed based on observations of the real world, describing things we see around us all day and every day. They are straightforward and have a certain elegant simplicity, and can be stated and understood without resorting to mathematics. While their implications and applications are vast, we'll confine ourselves to paddling.

If you answered in the affirmative to the questions above, feel free to skip this overview and move right into the heart of the book. If you need to get more familiar with the Laws, this brief review should get you up to speed. Even if you're a classical physics gunslinger, you might enjoy the historical context and original presentation of the 2nd Law.

The Laws of Motion

In 1665 Isaac Newton completed his undergraduate studies at Trinity College, Cambridge. The Great Plague had descended upon London, so Newton retired to his family's farm at Woolsthorpe Manor, Grantham. Over the ensuing year – what has been referred to as the "year of wonders" – Newton formulated new theories of physics and mathematics that underlie much of modern science and engineering. He showed that the movements of planets, moons, and other celestial objects, and the motion of things here on Earth, could be described in a single, unified framework. Newton did this using only algebra and geometry. Along the way, he developed differential calculus.

In 1687 Newton published his discoveries in the text *Philosophiæ Naturalis Principia Mathematica* ("Mathematical Theory of Natural Philosophy," commonly referred to as the *Principia*). The most well-known elements of this are his three Laws of Motion. We refer to them as laws because they're axiomatic. Newton saw them as self-evident truths that describe phenomena we routinely observe in the physical environment.

Everything that happens when we're in a hull can be described by Newton's laws of motion: why we float in the first place, stability, the various forces acting on the hull and paddle(s), why a hull moves when we paddle, the effect of trim, etc. So, as you read *The Science of Paddling*, know that you're both going old school (i.e., 17th century) and are in very, very good company.

This brief preface will review the three Laws of Motion and summarize how they relate to our favorite sport.

· · ·

The 1st Law states:[1]

> Every body perseveres in its state of being at rest or of moving uniformly straight forward, except insofar as it is compelled to change its state by forces impressed.

First, this means that an object ("body") like a canoe, kayak, or SUP will remain stationary unless you lift it up from the storage rack, or apply a paddling force when it is motionless atop quiet water. Stationary is one example of a hull's "state."

The other state is a hull moving forward at a constant (uniform) speed along a straight course.[2] From experience, we know that if we apply a forward paddling stroke, the hull will no longer move at a constant speed but will instead speed up. The force of the paddling stroke has "compelled" the hull to change its state from one speed to another. If we apply a corrective stroke, such as a draw or pry, we are compelling the hull to change its state from moving straight to turning. We also know from experience that a hull slows down when we are no longer "compelling" it. This is because drag forces – friction, form, and wave – are "impressed" upon the hull, constantly changing its state from one speed to another. Finally, changes in trim can induce asymmetric forces fore-and-aft that compel a hull to turn and then attain a new heading when acted upon by the dynamic pressure (in effect, a force) of current.

---

1 Newton's *Principla* was written in Latin, the 17th century's *lingua franca*. The translations of the laws of motion used here are from Bernard Cohen and Anne Whitman, *The Principia: Mathematical Principles of Natural Philosophy*, University of California Press, Oakland, California (1999).

2 A stationary hull has zero speed in any direction. This is a trivial special case of "moving uniformly." Newton may have anticipated folks would quibble if he didn't separately name this as the rest state.

As you read through these examples embodying the 1st Law, visualize each scenario: where the various forces act (are impressed) and how the hull responds (changes its state). Most likely, you have compelled a hull to change its state by forces impressed in many ways. You may not have referred to it in quite those terms.

· · ·

The 2nd Law states:

> A change in motion is proportional to the motive force impressed and takes place along the straight line in which that force is impressed.

Compare the 1st and 2nd Laws. In the 1st, Newton cites "force," while in the 2nd, he cites "motive force." This means we're talking about two different categories of force. Here motive force means impulsive force. Note also how the body in the 1st Law has a state, while in the 2nd, it has "motion." By motion, Newton meant *momentum*, which is the product of an object's mass and velocity. Consequently, a change in an applied impulse force, such as that delivered over a paddle stroke, changes the momentum of the hull/paddler system.

In Chapter 4, we'll see that momentum has a direction since it is the product of mass (a scalar, which is directionless) and velocity (which has a direction and therefore is a vector). For example, suppose you paddle out of a quiet shoreline eddy into a river's current, aiming straight for the far shore. The force exerted by the incoming current acts perpendicular to your intended direction of travel. This will move you downstream. The degree that your direction of travel is deflected depends on the dynamic force of the current vs. the force generated by your paddling strokes. We'll cover this in Chapters 6 and 9.

Since the change in impulsive force is proportional to the change in momentum, if you double the impulsive force, the resulting change in momentum also doubles; tripling leads to tripling, etc. The more impulse you generate by paddling strokes, the faster you go. We'll explore this in Chapter 7.

As we'll see in Chapter 2, after adopting a hull-centric reference frame, the wake behind a moving hull reflects a change in the incoming water's momentum. This change in momentum exerts itself as the *form drag* force on the hull.

Those of you who took high school physics, or studied engineering or science in college, likely learned that the 2nd Law is "force equals mass times acceleration." Why are these two statements apparently different? We'll cover that in the Extra Credit section.

· · ·

The 3rd Law states:

> To any action there is always an opposite and equal reaction; in other words, the actions of two bodies upon each other are always equal and always opposite in direction.

Here, "action" refers to force. Newton wrote the 3rd Law to encompass orbiting celestial bodies, billiard balls bouncing off each other, and objects in static equilibrium.

Newton's argument that the Earth and Moon are mutually attracted by gravitation forces was highly contentious when he proposed it. Critics claimed he had introduced invisible "occult forces" to make his mechanics work. When pressed as to what gravity was (as opposed to what it *did*), he famously wrote, "*Hypotheses non fingo*," or "I frame no hypothesis." His was an applied, empirical science and not merely philosophy. It would take a few centuries before Albert Einstein showed that objects having mass warp the fabric of space and time, thus creating the effect we call gravity. In the meantime, Newton's laws worked, and they still do.

We experience the 3rd Law when we place one of our hulls in the water. The hull floats. This is because the hull displaces a volume of water, leading to a buoyancy force that precisely equals the hull's weight. The buoyancy force acts upward, while the weight of the hull acts downward in the direction of gravity's pull. When we climb aboard our hull, it sinks lower, displacing a further volume of water whose weight equals our own. If we remain afloat, these opposing forces – buoyancy and weight – balance each other. Our weight is the "action," while the buoyant force is the "reaction." We'll dig into this more in Chapter 5 when investigating a hull's stability in roll.

You're also experiencing the 3rd Law while reading this. I'll assume you're sitting on a chair or couch. Your weight exerts a downward force due to gravity: an "action." The chair is exerting a reaction force in response to this action. These two forces precisely balance each other. Otherwise, your weight will cause the chair to collapse. At that point, you've entered the regime described by the 1st Law, moving downward until another force stops you. Looking around the room you are sitting in now, you may notice many objects sitting atop shelves, tables, the floor, etc. Each of these objects exerts a downward force equal to its weight. The shelves, tables, and floor beneath these produce reaction forces to keep things from falling through the floor. Newton's 3rd Law is all around us all day long.

Extra Credit

When Newton wrote the *Principia*, calculus hadn't been invented; he was inventing calculus as he went (as was Gottfried Leibniz in parallel). Newton's work was couched in algebra and geometry. His original statement of the 2nd Law doesn't use the words "rate of change" or "rate of change with respect to time." This is how we now describe acceleration in terms of velocity. These are concepts we take for granted; in the 17th century, they were hotly debated.[3] The rate of change is now commonly used because we have differential calculus.

Here's how you can derive the familiar relation between force, mass, and acceleration from Newton's original wording of the 2nd Law. First, in Chapter 7, you'll learn that an impulsive force – the product of force and the time interval over which it is applied – is related to a change in momentum via

---

3 See Amir Alexander, *Infinitesimal: How a Dangerous Mathematical Theory Shaped the World*, Scientific American / Farrar, Strauss and Girous, New York (2014).

$$F_p \Delta t = m \Delta v , \tag{0.1}$$

where F_p is the propulsive force, which for paddlers is the paddle reaction force in the direction of travel. For paddlers, the combined mass of hull, paddler(s), and gear is m; t is time, and v is velocity. Δt signifies an interval of time, while Δv indicates a change in velocity in response to the impulsive force. If we divide both sides of the equation above by Δt,

$$F_p = m \frac{\Delta v}{\Delta t} . \tag{0.2}$$

We'll now introduce a highly controversial concept from Newton's time: infinitesimals. Suppose we let the change in time Δt become infinitesimally small. In that case, the fraction on the right-hand side of this equation represents the instantaneous change in velocity over that infinitesimal period. In the limit as Δt approaches zero, this fraction represents the hull/paddler system's instantaneous acceleration a. So, in that limit,

$$F_p = ma . \tag{0.3}$$

This is the form of the 2nd Law we likely learned in school. Congratulations: you've just invented differential calculus! Now let's move on to paddling.

CHAPTER 1

Tandem and Solo

Introduction

We begin our journey with an observation and a question. Tandem canoes and kayaks are propelled by two paddlers, yet they aren't twice as fast as equivalent solo hulls. Why not?

Our hulls move through the water due to forces we apply through our paddles, with reaction forces transmitted through our bodies into the hull via our torso and legs. This motion embodies Newton's 3rd Law: For every action, there is an equal and opposite reaction. The water exerts a retarding force against the hull known as *drag* that is constantly trying to slow us down. While the details are rather complex, requiring that we solve vector nonlinear partial differential equations, drag is just a force that resists our efforts proportional to the square of the hull's speed. The quadratic dependence of drag upon speed, which holds for all canoes, kayaks, and SUPs regardless of their shape or the number of paddlers, will show us why tandem hulls aren't twice as fast as solos. And we'll only need algebra to prove it.

To begin our analysis, consider Fig. 1.1, which depicts a canoe moving from the upper right to lower left with a speed U. The hull has maximum beam width B and waterline length L. A drag force resists the hull's motion.

At very low speeds, the drag force experienced by a canoe moving through the water is proportional to the hull's speed times a friction coefficient. The drag force at other speeds is proportional to the product of the hull's *wetted area* – the surface area of the hull that's below the waterline, and hence is wet –and the square of the speed.

Since the beam width and below-water depth of any hull is a percentage of the waterline

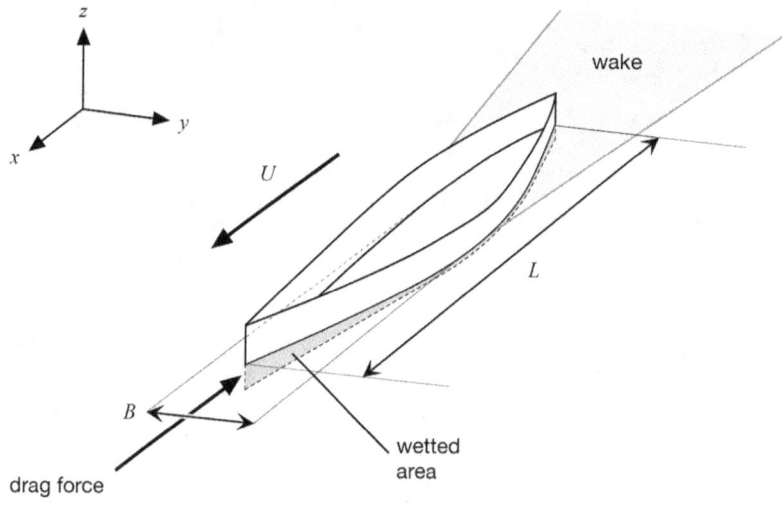

Figure 1.1: Hull moving over water.

length, we can assume the wetted area as proportional to[4] the square of the boat's length. All hull dimensions – beam, height at the stems, freeboard, etc. – can be expressed as a fraction of the length. This dimensional interdependence lets us do some rule-of-thumb modeling to develop insight into how hulls comparatively behave. And all it will take is algebra and a little physics.

Paddlers and Power

To compare tandem and solo hulls, let us assume that we have three well-matched paddlers, each capable of exerting the same paddling power. One of these paddlers is in a solo boat, while the other two are in a tandem of equal length that is hydrodynamically equivalent to the solo. The paddlers' power counteracts the drag forces on their respective hulls to provide forward speed.

Physics shows that the exerted paddling power P is proportional to the drag force times the speed. Recall that the drag force is proportional to the square of hull speed U relative to the water, and linearly proportional to the wetted area A. As a result, the exerted paddling power is

$$P \propto U^2 \times A \times U \propto U^3 L^2 \,, \tag{1.1}$$

where we have used the fact that the wetted area A is proportional to the square of the waterline length L; the truncated infinity sign means "proportional to." Thus, paddling power is proportional to the speed cubed times the length squared. Equation (1.1) yields an expression for the speed U in terms of power and length,

---

4 Note that we'll use "proportional to" rather than "equal to" throughout this chapter. We're deriving rules of thumb to develop an intuitive and broadly applicable understanding, rather than solving the Navier-Stokes equation for the detailed fluid mechanical behavior of flow around a specific hull.

$$U \propto \sqrt[3]{\frac{P}{L^2}} = \sqrt[3]{P} \times L^{-\frac{2}{3}} \cdot \qquad (1.2)$$

Consequently, the hull speed is proportional to the cube root of the paddler's applied power.

Now for our tandem, which has two paddlers each capable of exerting the same paddling power P, we can write the relationship between paddling power and speed as

$$2P \propto U^3 \times L^2 \; . \qquad (1.3)$$

since there are two paddlers, there is twice the power in the tandem hull compared to the solo. Solving equation (1.3) for the tandem's speed U,

$$U \propto \sqrt[3]{\frac{2P}{L^2}} = \sqrt[3]{2P} \times L^{-\frac{2}{3}} = \sqrt[3]{2} \times \sqrt[3]{P} \times L^{-\frac{2}{3}} \; . \qquad (1.4)$$

Since we have assumed that the two hulls are the same length L, we see that subject to the assumptions the tandem will be faster than the solo by a factor of the cube root of two, or 1.26, e.g., about 26%. So to a *first approximation*, if the solo cruises at 6.0 mph (2.68 m/s), then an equivalent tandem could cruise at 7.5 mph (3.35 m/s), *all other factors being equal*. We see that a tandem is not twice as fast as a solo despite having twice the power plant because of the nonlinear relation between paddling power and speed. This difference – less than twice as fast – is generally consistent with what we see on the water.

Role of Paddler Weight

Our model neglects the effect of the tandem's greater weight due to its second paddler. Since the tandem will weigh about twice the solo – for most hulls, the weight of the paddlers and their gear is the dominant component of the combined weight – it will sink deeper in the water than the equivalent solo hull. This increases the tandem hull's wetted area compared to the solo, which should slow the tandem down in comparison due to increased drag.

When you think about it, paddling is just moving a hole through the water. This hole is created from the water that the hull displaces. The total weight of the paddler(s), plus the weight of their hull and gear, equals the weight of the water displaced by the loaded hull. This is *Archimedes' Principle* ("Eureka!") – assuming, of course, that you're afloat. If the weight of the paddler(s) increases, the amount of water displaced increases.

The weight of the displaced water equals the density of water times the displaced volume of water. The displaced volume is proportional to the product of waterline length, the beam, and the submerged depth, a *cubic* dimension. Since the beam and depth can be expressed as a percentage of the waterline length, we see that weight is proportional to the length of the hull cubed.

As you paddle your "moving hole," water is continually rushing in behind the hull to fill the void it just left behind. This moving hole is why stern wake riding works: you are essentially pulling along a wave train of water as you paddle, which a trailing canoe positioned close to the stern can sit in. The mass of this moving hole increases in proportion to the amount of water displaced by the hull, which is linearly proportional to the weight of the paddler(s) because of Archimedes' Principle.

Recall that the drag force on a hull is proportional to the wetted area times the square of the hull speed. Since the wetted area is proportional to the waterline length squared, and the cube of the length is proportional to the weight W of the displaced water, then

$$\text{Wetted Area} \propto L^2 \propto W^{\frac{2}{3}} . \tag{1.5}$$

Thus, we can express the relationship between power and speed in terms of the paddler(s) weight as

$$P \propto W^{\frac{2}{3}} \times U^3 , \tag{1.6}$$

which, solving for the speed U yields

$$U \propto \sqrt[3]{P \times W^{-\frac{2}{3}}} = \sqrt[3]{P} \times W^{-\frac{2}{9}} . \tag{1.7}$$

So for a fixed amount of paddling power, hull speed goes as the minus two-ninths power of weight and the cubed root of paddler power. Not exactly an intuitive relationship, eh? But plugging in a few numbers will provide some insight into the significance of each effect. For example, if a paddler gains 5% of their weight eating cookies and eggnog over the Winter months, but doesn't lose any paddling power, then their cruising speed will decrease by

$$\left(1.05\right)^{-\frac{2}{9}} = 0.99 ,$$

a speed loss of 1 percent. If that same paddler did a tad too much carbo-loading and instead gains 10% over their ideal weight, then their cruising speed will decrease to

$$\left(1.1\right)^{-\frac{2}{9}} = 0.98 ,$$

a drop of 2 percent. That may not seem like much, but a 2% decrease in cruising speed means that if our paddler is a racer, a course they paddled last year in 1 hour will now take 1:01:13, or one minute and thirteen seconds slower. In paddlesport racing, this can equal or exceed the margin of victory.

Fig. 1.2 shows the impact on the finishing time of a hypothetical 1-hour long paddling race for increasing paddling power, and decreasing paddler weight.

Returning to our tandem versus solo paddling analysis, let us now assume that all three of our well-matched paddlers not only have the same paddling power, but they all weigh the same, too. For simplicity, we'll again assume that the paddlers' weight is significantly larger than the weight of the hulls. The tandem hull plus its paddlers will then weigh about twice the solo and its paddler. This 100% increase in weight will decrease the tandem's cruising speed to

$$(2)^{-\frac{2}{9}} = 0.86 \; \cdot$$

Our initial analysis showed that the tandem's double power plant would make it about 26% faster than an equivalent solo and paddler. Because of the difference in combined weight between tandem and solo, the tandem hull's increased wetted area will decrease its power advantage by 1.26 × 0.86 = 1.08, e.g., about an 8% advantage in speed over a solo. If a solo paddler can traverse a course in an hour, an equivalent tandem could paddle the same course in about 0:55:30, assuming all other factors are equal; if a solo cruises at 6 *mph* (2.68 *m/s*), then a tandem with power-equivalent paddlers would cruise at about 6.5 *mph* (2.91 *m/s*). This result, which reflects the impact of a tandem's increased drag due to its increased wetted area, is far more reasonable in light of real-world results than the 26% difference calculated from the difference in power plant alone.

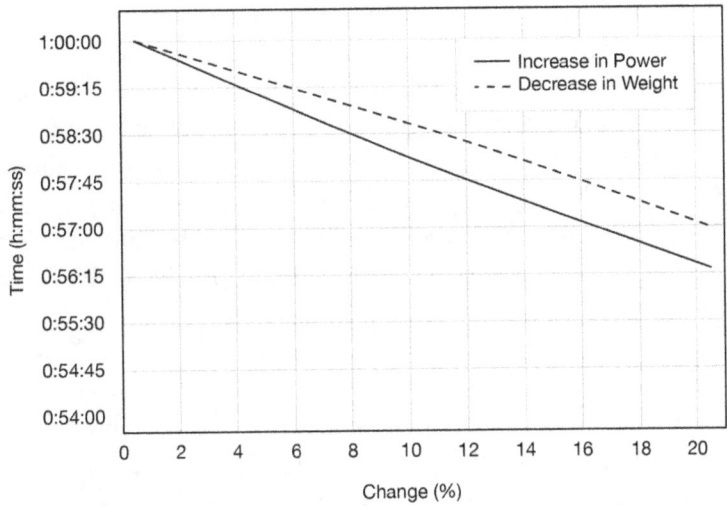

Figure 1.2: Impact of Power and Weight on Race Time.

The question now is, should a performance-oriented paddler focus on gaining paddling power, or losing weight? We see from equation (1.7) that cruising speed is related to a type of power-to-weight ratio:

$$U \propto P^{\frac{1}{3}} \times W^{-\frac{2}{9}} = \sqrt[9]{\frac{P^3}{W^2}} \; \cdot \qquad (1.8)$$

The expression inside the root shows that increasing paddling power P by appropriate training has more impact than decreasing a proportional amount of weight W because of the cube and square in the fraction, respectively. Developing paddling-specific power will pay off a bit faster than losing weight. And paddlers with an excellent power-to-weight ratio will perform the best of all. That's why many female paddlers, as well as light mixed teams, do so well at the races.

Another way to look at it is, should I lose weight or trade in my standard-weight hull for an ultra-light hull? Wouldn't the effect be the same? Here, we can do some simple financial modeling. Say the cost of a health club membership is around $10 US per week. One can sustainably lose about 1.25 pounds per week through proper diet and exercise. If you need to drop 10 pounds sustainably, you can do this in about eight weeks. At $10 US per week this leads to about $80 US in health club membership fees. Compare that to the $500+ US you'll pay as the premium for an ultra-light hull, and you'll see that going to the local health club to lose weight is more cost-effective... and may increase your power to weight ratio to boot!

Aside from losing weight, are there other ways we might reduce the wetted area of a hull to reduce drag and increase speed? How about adding a helium-filled balloon to create more buoyancy? A quick review of various racing society rules showed that there was nothing explicitly prohibiting balloons as long as they are securely attached to the hull.

Helium provides about 1 gram of buoyancy – a force to counteract gravity – per liter. A 12" (30.5 *cm*) diameter balloon provides about 14 grams of lift per balloon. If a paddler weighs 165 *lb.* (about 75 *kg*), to offset this weight one would need 5,357 of these helium balloons. Obviously, this sizable collection won't fit inside any hull, but it might be fun to watch someone try to assemble it. To "lose" a quick 10 *lb.* (4.5 *kg*) using helium balloons one would need 325 of them. To "lose" a quick 3.3 *lb* (1.5 *kg*, or 2% of the paddler's weight) one would need 107 helium balloons. This could provide a small but measurable speed benefit as long as the balloons didn't increase wind drag. Plus it might be fun, which ultimately is the point, isn't it?

Take-Aways

- Tandem hulls are not twice as fast as equivalent solo hulls because speed varies as the cube root of the combined paddling power of all paddlers in each hull.
- The drag force that slows our hulls is proportional to the speed squared at all but the slowest speeds.
- The combined weight of the hull and its contents increases the hull's wetted area, thus increasing drag. All other things being equal, a lighter hull and lighter paddlers will be faster. Or, for a given speed, a lighter hull with lighter paddlers requires less energy to paddle.

Extra Credit: The Drag Equation

Where does this squared dependence of drag on speed come from? We can answer that question using a nifty tool called *dimensional analysis* to derive the so-called Drag Equation.

When I first learned about dimensional analysis as an undergrad, I thought it was magic. In essence, you list all of a problem's relevant parameters, toss them in a bag, shake, and an equation pops out that describes mathematically what you are interested in. This is a gross oversimplification, and while I understand the mathematical foundation that underlies dimensional analysis and know why it works, it is still almost magical.

In order to parameterize the forces acting upon a moving hull we'll propose the following as the relevant factors:

- The speed U of either the fluid or the hull
- The fluid's kinematic viscosity υ, which characterizes fluid friction
- The fluid's density ρ, i.e., its mass per unit volume
- A characteristic area of the hull A, such as its below-water wetted area
- The drag force F_D.

The hull speed is a logical choice since we wish to determine the drag force's dependence on speed. The kinematic viscosity lets us investigate the impact of fluid friction due to motion. The fluid's mass, conveniently represented via its density, will play a role in inertial effects. The area is relevant because we expect one or more hull dimensions will play a role in drag. All of these parameters are either known or measurable. And finally, the drag force is included since we wish to determine how it relates to the other parameters.

Dimensional analysis proceeds by positing an equation expressed in terms of a function f of all these parameters:

$$f\left(U, \upsilon, \rho, A, F_D\right) = 0 . \tag{1.9}$$

Equation (1.9) may seem pretty abstract; don't worry, it is. The key insight is to recognize that the right-hand side of equation (1.9) – the number zero – is *dimensionless*. All of the parameters above have dimensional units, such as meters per second or kilograms per cubic meter. But zero has no units; it is nondimensional. Therefore, the function f must also be dimensionless (or if you prefer, free of dimensional units), and comprise a few nondimensional quantities. We construct these nondimensional quantities from the parameter list.

Fortunately, there are only two combinations of the parameters that are nondimensional. One grouping, well known in fluid mechanics, is the *Reynolds Number R_e*. The Reynolds Number is defined in terms of a characteristic length L as

$$R_e \equiv \frac{UL}{\nu} = \frac{U\sqrt{A}}{\nu} . \tag{1.10}$$

For convenience, we've assumed that the characteristic length L is expressed as the square root of the characteristic area A. Don't worry what the area A or length L are right now; it all works out to within a multiplicative constant at the end.[5]

The Reynolds Number is a dimensionless quantity, which meets our requirements. The numerator includes the speed U, while the denominator includes the kinematic viscosity υ. Consequently, the Reynolds Number is a ratio of inertial (e.g., motion) and viscous effects; a large R_e indicates inertial effects dominate viscous effects, and vice versa.

The other relevant nondimensional quantity that pops out from our list of parameters

---

5 Or think of it as part of the magic of dimensional analysis. Whatever works for you.

is the ratio of the drag force F_D to the product of the flow's density, speed squared, and characteristic area A:

$$\frac{F_D}{\frac{1}{2}\rho U^2 A}. \tag{1.11}$$

The product of density and speed squared has units of pressure. When we multiply this product by one half the result is called the flow's *dynamic pressure*, or the pressure exerted over a surface by a fluid in motion. Multiplying the dynamic pressure by the characteristic area A yields the dynamic force exerted on the surface, since force is pressure times the area over which it is exerted.[6]

The nondimensional ratio of forces (1.11), along with the Reynolds Number (1.10), make our abstract function f a bit less abstract:

$$f\left(\frac{F_D}{\frac{1}{2}\rho U^2 A}, R_e\right) = 0. \tag{1.12}$$

Because the only unknown in this equation is the drag force, we can rewrite equation (1.12) as

$$\frac{F_D}{\frac{1}{2}\rho U^2 A} = r\left(R_e\right), \tag{1.13}$$

or

$$F_D = C_D U^2 \tag{1.14}$$

for

$$C_D \equiv \frac{1}{2}\rho A r\left(R_e\right) \tag{1.15}$$

where we have defined C_D as a dimensional hull drag coefficient, expressed in terms of a new function r of the Reynolds Number R_e. Equation (1.14) is the Drag Equation. Dimensional analysis shows that the drag force varies as the square of the flow speed U, subject to very few assumptions.

What exactly is the function r? It's not that important unless you want precise answers for a specific hull. Yeah, fluid mechanics is like that. And that's OK. Our goal was to develop insight, not analyze a particular canoe, kayak, or SUP. Over the range of speeds we paddle, the Reynolds Number only varies so much, so the function r – which depends upon

---

6 Also, note that pressure is an isotropic quantity, meaning it has no preferred direction. When pressure acts upon a surface, such as the inside of a car tire or the surface of a kayak hull, it does so as a normal force. Force has a direction. Pressure-induced forces act perpendicular to the surface they adjoin.

the Reynolds Number – won't change with speed to a first approximation. This is borne out by experiment, as we'll see in Chapter 11. The weak dependence on speed also implies that we can reasonably assume the drag coefficient C_D is a constant for a given paddled hull.

Further Reading

Wen-Hsiung Li and Sau-Hai Lam, *Principles of Fluid Mechanics*, Addison-Wesley (1964).

Ain A. Sonin, "The Physical Basis of Dimensional Analysis," Department of Mechanical Engineering, MIT (2001).

CHAPTER 2

Kind of a Drag

Introduction

I curiously watched a tandem canoe paddler while I tried to wipe the sleep from my eyes on Memorial Day morning at Lakefront Park. He was alternately spraying a clear liquid on the bottom of his high-end racing hull and then frantically buffing it. Perhaps he was practicing some new warm-up exercise before the annual General Clinton 70-mile canoe race?

I learned instead that he hoped to make his boat faster by coating it with this magical liquid.

I asked about all the scratches on the hull; perhaps they might slow him down a bit? He admitted that there wasn't much he could do about those on race morning. But he fervently believed that the magic goo he was applying to the bottom of his canoe would more than compensate for the scratches.

Back in the day, I spent a couple of years of graduate school conducting research in a water tunnel,[7] exploring novel techniques for drag reduction on seagoing vessels. And I know that in some instances continuously injecting liquid polymers or air bubbles into the flow adjacent to a hull can reduce drag – more on that later. But these exotic techniques exploit physics that are far different than the more mundane problem of friction drag, the problem our tandem paddler hoped his hull polish would somehow overcome. Hydrodynamics – the study of fluid flow, and its effect on hulls and other objects – shows that the pertinent

---

7 Think wind tunnel but filled with water instead.

question was how much our tandem paddler and his partner would be slowed down by all of those hull scratches. Or how the very presence of their hull in water inherently creates more dominant drag forces unrelated to friction.

The drag forces on a canoe, kayak, or SUP are divided into three parts: *friction drag, form drag,* and *wave drag.* Whenever we paddle all three come into play simultaneously, and in varying degrees depending on speed and hull design. In this chapter, we'll look at the first two components. We will review wave drag in Chapter 3. Let's begin our investigation with friction drag.

The Rough Stuff

The most significant misconception people have about waxes, polishes, and other coating treatments is that they somehow allow their hulls to slip through the water faster and thereby reduce or eliminate friction drag. The hope is that water won't "stick" to the hull if coated this way. Nothing could be further from the truth. There are *other* possible benefits from polishes, and we'll consider those later in the chapter. But in terms of making a boat more "slippery" in the water, well, not so much. To understand why, let's look at what happens as water flows over our canoes, kayaks, dragon boats, and SUPs.

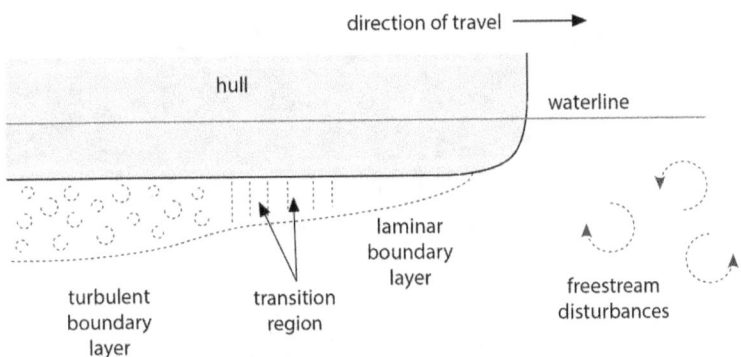

Figure 2.1: *Water flow beneath a hull (not to scale).*

Consider the flow of water along a hull, as depicted in Fig. 2.1. In this figure the boat is moving from left to right in quiet water. We'll assume that the boat is up to speed and traveling at a steady pace. We'll also adopt a paddler's frame of reference, where we think of the water as flowing past the hull rather than the hull past the water. This change of reference, called a Galilean transformation,[8] lets us move between frame of reference that differ by a constant relative velocity. The physics are easier to understand this way.

As water approaches the hull its constituent molecules, loosely bonded to each other by cohesive molecular forces,[9] are free to move anywhere they want as a liquid. When first encountering the boat, something interesting happens: the water molecules that contact the

---

8 Named after Galileo Galilei but attributed to Isaac Newton. Go figure.

9 If they weren't attracted to each other water as we know it wouldn't' exist —over time the water molecules would move apart due to Brownian motion. As a result, we likely wouldn't exist, either.

hull stick to it. Right at the hull surface, the water molecules and the hull move together. Since we've adopted a hull-centric frame of reference, the water right at the hull's surface has zero velocity with respect to the boat and paddler. This is called the *no-slip condition* in fluid mechanics. The no-slip condition holds for hulls made of Kevlar, carbon fiber, wood, fiberglass, aluminum, plastic, or anything else for that matter. And the no-slip condition holds whether the hull has been coated with liquid goo, a UV protectant, carnauba wax, or slime from the last paddling session. In all these cases the relative speed between the hull, and the water immediately in contact with the hull, is zero.[10]

Now we know that water flows past the hull as we paddle; if *all* of it stuck – the water both adjacent to the hull and extending away from it – we wouldn't be able to move. What happens to the water that isn't in contact with the hull, but is instead close to it?

"Close" is where the concept of a *boundary layer* comes into play. Boundary layers are the source of *friction drag* that continually works to slow us down.

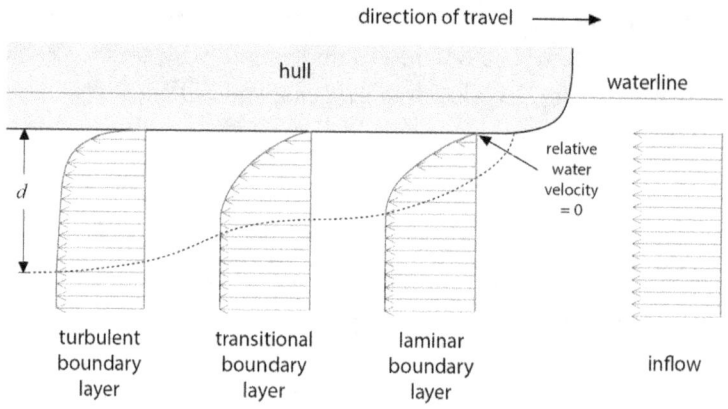

Figure 2.2: Boundary Layers (not to scale).

The no-slip condition requires the water's relative speed at the hull to be zero. From our Galilean transformation, the so-called freestream speed of water far away from the hull will equal the hull speed. Consequently, there is a variation or *gradient* in the water's relative velocity as we move away from the hull's surface, until we reach the point where the water's velocity equals the freestream speed. The region where this gradient exists is the boundary layer.

Boundary layer shapes are illustrated conceptually in Fig. 2.2 for laminar, transitional, and turbulent flow. The thin horizontal arrows denote the average flow speeds at various distances from the hull, where a longer arrow means faster flow speed. The dotted line denotes the thickness of the boundary layer where (for example) the flow speed equals 99%

---

10 Some have pointed out that water beads and rolls off suitably treated hull materials. But this is a different fluid mechanics phenomena, pertaining only to beads of water in air on a free solid surface. The surface treatment changes the inter-molecular forces on the water droplets' surface, making them adhere to each other more tightly and thus "bead." These treatments change the solid surface from *wetting* to *non-wetting* – for beads of water, in air. Different physics.

of the freestream speed. The boundary layer grows in thickness from the bow to the stern, for reasons we'll explore below. And as suggested by the different boundary layer profiles in the figure – laminar, transitional, and turbulent – the boundary layer's internal character changes as well.

A *laminar* boundary layer develops near the front of the boat. The water impinging on the hull travels uniformly in this case. If we injected colored dyes in the water just upstream of the bow we'd see that the dye moves along the hull there in nice, parallel lines, hence the term laminar – like laminations. The laminar boundary layer depicted in Fig. 2.2 has a gradual gradient of the water's speed as we move away from the hull. The slope of this gradient is indicative of the frictional forces on the hull. The more gradual gradient in the laminar case reflects lower wall shear stresses, which means less energy loss and lower frictional drag. If the flow over the entire hull were laminar this would reduce friction drag, and we'd go a bit faster.

Laminar boundary layers are sensitive beasts. Think of a laminar boundary layer like a little channel of water adjoining the hull. Inside this channel there is smooth-flowing water and tiny pressure oscillations. These pressure waves – called *Tollmien-Schlichting* or "TS" waves – travel inside this laminar boundary layer channel in the direction of flow, from the front of the hull towards the back, much like sound travels down a long hallway. TS waves are initiated by small disturbances in the upstream flow, such as those we encounter paddling in swirly or turbulent rivers or when closely following another boat.

The TS waves differ from ordinary sound waves in one significant aspect: they grow in amplitude as they travel along the hull. After travelling a certain distance, TS waves grow in amplitude enough to influence and then disrupt the laminar boundary layer's orderly structure. The nice parallel streamlines of the laminar flow start to wiggle and distort under these growing pressure waves' influence and become disorganized. The boundary layer starts its transition from laminar to turbulent. This *transition region* is shown in Fig. 2.1.

The hull speed, along with the viscosity of water, specify the location along the hull where transition occurs. If we represent the hull speed over the water as U, and the kinematic viscosity of water as ν, and the distance x from the front edge of the hull, then the location at which the flow becomes turbulent is determined by the following approximation:

$$\frac{Ux}{\nu} \approx 1 \times 10^6 . \tag{2.1}$$

The fraction on the equation's left-hand side is called the *Reynolds Number* (see Equation 1.10), usually written as "R_e." The Reynolds Number represents the ratio of inertial (momentum) forces to viscous (friction) forces in fluid flow, as we noted in Chapter 1. When the Reynolds Number equals approximately one million the flow over the hull becomes turbulent. Since the viscosity is a property of the water and 10^6 is a constant, equation (2.1) shows that transition to turbulence occurs closer to the bow at higher speeds and further back at lower speeds since the speed and transition location vary inversely for a fixed Reynolds Number.

An 18'6"-long hull traveling at 6.3 *mph* (2.8 *m/s*) in 68F (20C) water sees transition occurs a mere 14 inches (35.4 *cm*) from the bow. The length of the transition region will be even shorter. For this hull and speed, we see that the flow over most of its length is turbulent.

If the upstream disturbances in the flow are large, perhaps due to whitewater eddies or the wake of a boat we are following, then transition will occur even closer to the bow.

Once the flow is turbulent there are no longer any nice, orderly streamlines inside the boundary layer. Instead, the flow there becomes chaotic and highly three-dimensional. The flow visualization in Fig. 2.3 illustrates this. In this plan view we're looking at a portion of the hull's bottom, with water flowing from right to left. A dye has been injected into the water upstream at several points spaced across the hull's width. Notice how the dye streaks change from nice parallel lines to a chaotic structure in the flow's transition to turbulence. Note also how the dye steaks oscillate in the transition region just upstream of the prominent turbulent burst "cloud."

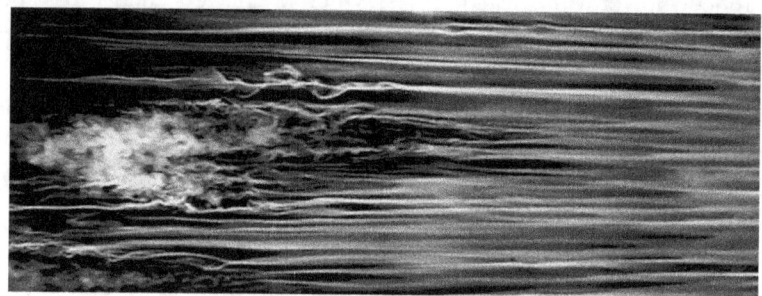

Figure 2.3: Flow visualization: Plan view of a boundary layer becoming turbulent.

The chaotic nature of the flow grows downstream of this area as the boundary layer becomes fully turbulent. The gradient of the averaged velocity gradient inside the now turbulent boundary layer becomes more pronounced, more "full," than in the laminar case, as depicted in Fig. 2.2. The high-speed outer flow now mixes with the low-speed flow near the wall, increasing its momentum. The steeper flow speed gradient generates the increased frictional forces arising from shear stresses on the hull's surface, which are larger than in laminar flow. The boundary layer itself becomes thicker as well, as we are putting more of our precious paddling energy into generating turbulence.

There are ways to extend the hull's laminar flow region. But these techniques entail impractical measures like generating out-of-phase sound waves ahead of the hull (as demonstrated by my grad school office mate Chuck Gedney in his Ph.D. thesis), or selectively heating the hull surface (as demonstrated by Dan Nosenchuck in his Ph.D. thesis at Cal Tech) to counteract the TS waves' growth. These techniques delay transition – i.e., move transition toward the stern – by about 50% under pristine laboratory conditions. When it comes to turbulence and friction drag, we can't win, we can't break even, and we can't get out of the game.

Now the no-slip condition, boundary layer transition, and turbulence are going to be there whether we coat our hull or not. But are there any things we can do to make our boat go a little faster? Aside from buying a faster hull shape – and more training! – the first thing to consider is the condition of the hull. Many of us who paddle on shallow rivers have had one or more close encounters with logs and rocks. The resulting scratches are a concern. But should they be? Are some scratches worse than others?

Most minor imperfections in a hull's smoothness will slightly increase viscous drag, on the order of a couple of percent. At the other extreme, a group of students at the University of Michigan took an aluminum canoe, bludgeoned it with sledgehammers all along its length, and measured the resulting increase in drag force in a towing tank. The amply distorted hull exhibited a 6% increase in total drag, so how bad can *our* hull be?

Keep in mind viscous drag is only one component of the total drag force exerted on a hull when paddled, as shown in Fig. 2.4. As we approach the so-called hull speed, viscous drag will be overshadowed by wave drag effects that we'll explore in the next chapter, plus another force called form drag. Form drag is a property of the hull's shape, while wave drag arises from the fact that we're paddling along the interface between water and air. So, addressing a 1% increase in viscous drag by polishing a hull will result in at most a comparable decrease in total drag, and will be most noticeable at the lowest speeds.

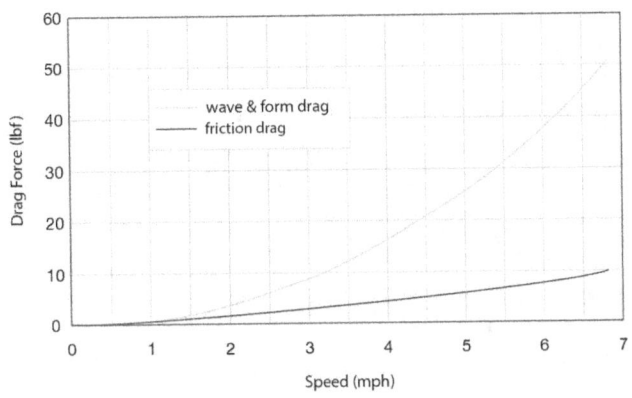

Figure 2.4: Example of drag forces vs. speed.

But how rough is rough? If a hull is "smooth enough," will it slip through as if it were perfectly smooth? And is there a rule of thumb regarding roughness for those of us who obsess about our hulls? Well, of course!

Even though the flow over most of our hulls is turbulent, there is an exceedingly thin layer of the flow immediately adjacent to the hull's surface that is laminar, no matter what. This *laminar sublayer* is far thinner than the boundary layer in which it resides. The laminar sublayer exists because there is a difference between molecular diffusion and the propagation of vorticity – and vorticity propagation feeds turbulence in the outer boundary layer. In other words, serious techno babble. The sublayer is the key to understanding how surface roughness impacts friction drag.

Like the laminar portion of the boundary layer itself, the laminar sublayer is easily disturbed. Hull roughness can destabilize the sublayer, leading to increased turbulence through a process called *bursting* where microscopic jets of water launch away from the hull into the main boundary layer, further energizing the turbulence there and increasing drag. But as long as the height of any surface roughness is *small relative to the thickness of the laminar sublayer*, the hull will behave as if smooth; the sublayer will continue to be stable and laminar.

The laminar sublayer thickness, represented by δ_1, is approximated by

$$\frac{\delta_1}{x} = 103 \times \left(\mathrm{Re}_x\right)^{-\frac{9}{10}}, \tag{2.2}$$

where Re_x is the Reynolds Number cited above, and the length scale is the distance x from the bow. Fig. 2.5 is a plot of the sublayer thickness along the hull for an 18.5-foot hull running at 6.3 *mph* (2.8 *m/s*) in 68F (20C) freshwater. Note that the sublayer thickness is inversely proportional to a power of hull speed U, which means the sublayer is thinner the faster we go. A thinner sublayer exacts a more demanding limit on hull smoothness: the faster we go, the smoother our hull needs to be to keep the sublayer stable.

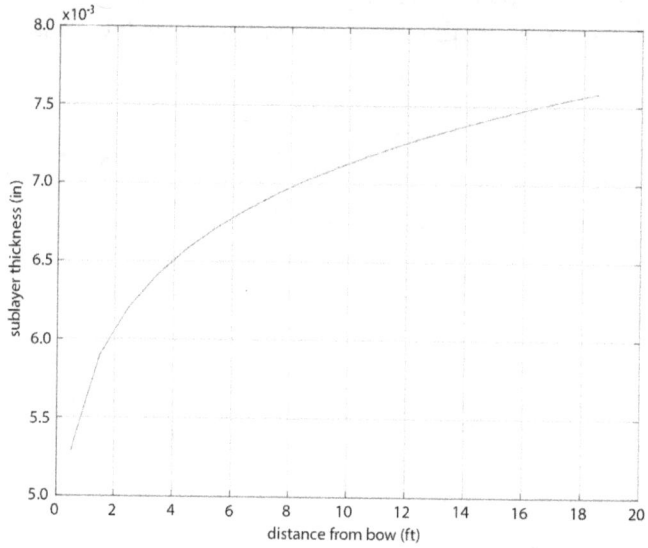

Figure 2.5: Sublayer thickness (mils) vs. distance along the hull (ft).

Fortunately for us paddlers, we don't move all that fast, at least compared to a car or a jet. As shown in the plot, our hull's sublayer thickness at 6.3 *mph* varies from about 5 thousandths of an inch near the bow to around 8 thousandths of an inch near the stern. So, to ensure that the hull is hydrodynamically smooth, make sure it is no rougher than about 1 to 1.6 thousandths of an inch average roughness height, or about one-fifth of the sublayer thickness. In more familiar terms, 1.4 thousandths of an inch is about the average grain size of 320 grit sandpaper. If the hull is as smooth as that, as far as the water is concerned, it is perfectly smooth. Roughness greater than that will introduce more turbulence into the boundary layer, with an increase in frictional drag.

So, where should we focus our energy and resources to reduce friction drag, if anywhere? Certainly, a scratched and rough hull will foul more easily. Deep scratches should be repaired if for no other reason than to prevent moisture penetration into the underlying fibers of a Kevlar, fiberglass, or carbon fiber composite lay-up, or the wood of a wood strip or wood-canvas layup. Those are reasons enough to keep a hull in good repair. And if we want to do some polishing, *and the polish fills or grinds down any superficial scratches*, there will be a *small* decrease in frictional drag at paddling speeds. Polishing will have the greatest impact over the first few feet of the hull, since small surface imperfections there can help destabilize a laminar boundary layer and lead to an even earlier transition to turbulence. So, polish the bow and the front meter or so of the hull, and keep the rest smoother than about 0.002".

And as to employing liquid goo, the chief benefactor there may lie between our ears. If we feel that it makes us paddle faster, we probably will.

Now a completist will ask if there is a component to drag that *doesn't* depend on friction. It turns out that there are two. We'll consider one of these – form drag – in the next section.

Waking Up

Straighten your arm out in front of you, with your hand flat and parallel to the floor. Now briskly move your hand side-to-side, arm straight. Notice the sensations of air passing it and any resistance to your motion.

Next, repeat this experiment with your hand still flat, but now held vertically (e.g., perpendicular to the floor), briskly moving your hand side-to-side, arm straight. Notice anything different? Do you feel more force distributed over your hand in this orientation than before?

This simple experiment is even more dramatic if you stick your arm out the window of a moving car.[11] With your hand held horizontal the air rushes past it. With your hand held perpendicular to the air flow the amount of force on your hand can be quite large.

Why are the forces you experience in these two cases different? If you said, "friction," note that friction drag force is proportional to the object's surface area – here, the surface area of your hand. The surface area of your hand doesn't change whether you hold it vertically or horizontally. While friction drag is present in both cases, there's something else going on.

The simple answer is *form drag.*

Form drag arises from the mere presence of an object – i.e., a form – in a moving fluid. When we're in our canoe, kayak, dragon boat, or SUP we notice a drag force difference if we try to move the hull straight ahead versus trying to side-slip it. We all know that stream-lined hulls move briskly traveling straight ahead, akin to when your hand's cross-section is aligned with air flowing past it as in the thought experiment above. But when we draw a hull 90-degrees to the side, it's like trying to slide a barn door through the water. The difference between these two cases suggests that form drag depends on the hull's shape, or rather the shape of the hull presented to the flow. So, what is form drag, and what causes it?

Say Hi to Mr. Bernoulli

Since form drag is a different phenomenon than friction drag, we might consider modeling it using so-called *inviscid* or "friction-free" fluid mechanics. Some readers may be familiar with Bernoulli's equation, which describes an inviscid flow in terms of fluid pressure and flow speed. A straightforward Bernoulli analysis would proceed as follows. Consider the plan view of a hull, as shown in Fig 2.6.

---

11 If not dangerous at high speeds, so be careful out there!

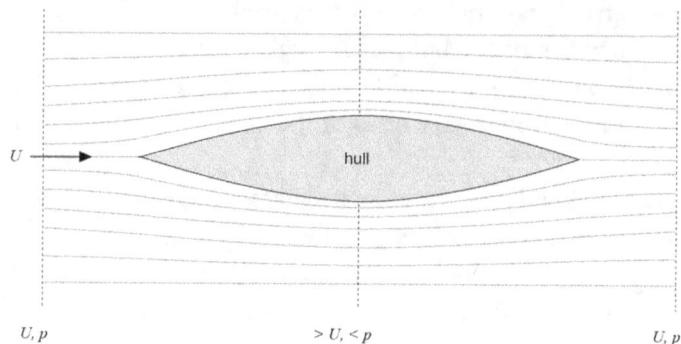

U, p $> U, < p$ U, p

Figure 2.6: Plan view of inviscid flow around a hull.

A uniform flow moving left-to-right with speed U impinges on the hull. The hull's keel aligns with the flow direction. A series of streamlines flank the hull's port (left) and starboard (right) sides. Streamlines are curves that are tangent to the local fluid flow direction; they are "aligned with the stream." Some distance away from the hull the streamlines are parallel since the hull's presence is not felt there. But closer to the hull, the streamlines are squeezed together more and more as the flow passes its widest point (corresponding to the hull's beam), then spread apart downstream.

In a Bernoulli analysis there can't be any shear forces along the hull, or behind the stern – remember, this is a friction-free model – hence the flow that divided into two halves at the bow rejoins behind the boat. The flow must speed up as it approaches the widest point of the hull, then slow down afterwards until it once again equals the flow's freestream speed U. Everything matches; the hull merely displaces the water. Also, no water is lost or created, a consequence of *conservation of mass*.

Bernoulli's Principle requires that the total energy in the flow must be conserved. In this friction-free model, there are two energy mechanisms: kinetic energy due to the water's motion and potential energy in the form of pressure that can perform work. As the figure suggests, the uniform flow upstream and downstream of the hull both have the same flow speed U and pressure p. Since the flow speeds up midships, the kinetic energy is greater there than at the bow. And because energy is conserved, the potential energy must decrease at midships as well.

Most of us have an intuitive idea of what pressure is. But we often confuse force – which has a direction and thus is a *vector* quantity – with pressure. Pressure is *isotropic*, which means it has no direction inside the fluid, just magnitude. If you have ever inflated a tire or a basketball, you know that pressure pushes the walls of these flexible objects outward. Once inflated, neither tires nor basketballs fly away; there are no unbalanced forces which would cause them to accelerate. The walls of the tire and basketball, subject to internal air pressure, achieve an equilibrium. The inflated tire or basketball are always exerting an inward-directed elastic force to maintain their shape. Newton's 3rd Law requires that all balanced forces have equal magnitude and opposite direction. The air pressure inside is exerting an outward-directed force against these objects' walls to balance the competing elastic force; the atmospheric pressure on the outside is doing the same thing, directed inward. Pressure, then, acts as a distributed force *on surfaces*, directed perpendicular to the surface at every point in contact with the fluid.

For our hull, this means that at the bow and stern, where the flow is nearly equal to the freestream flow, there are equal pressures and thus equal forces that press against the hull. These are greater than the pressures that act amidships because of the Bernoulli principle: there must be a reduced pressure where there is a faster flow speed. The pressure distribution can be summed over the hull's surface area to determine the net force in the x-y plane since force is the product of pressure and area, directed perpendicular to the hull at every point. Any net force will be either a propulsive force or a drag force, or a lateral force if the body is not symmetric.

Performing this calculation for our inviscid flow around the hull leads to a net force of zero. The process for doing this calculation is illustrated in Fig. 2.7. This figure is a plan view of just the bow and stern of our hull from Fig. 2.6; including the rest would make the illustration a bit too busy, while the analysis is the same. Exemplary forces due to fluid pressure are depicted as vectors with magnitude F, perpendicular to the hull surface, at four numbered locations. These are a subset of the pressure-induced forces from the entire pressure field around the hull. Since the hull is symmetric about the keel line the pressure forces at these symmetric fore and aft positions are equal.[12] Each force vector is depicted along with its respective components in the x- and y-coordinate directions. We decompose the vector into perpendicular components to deduce the net force in the direction of travel, and any net side force.

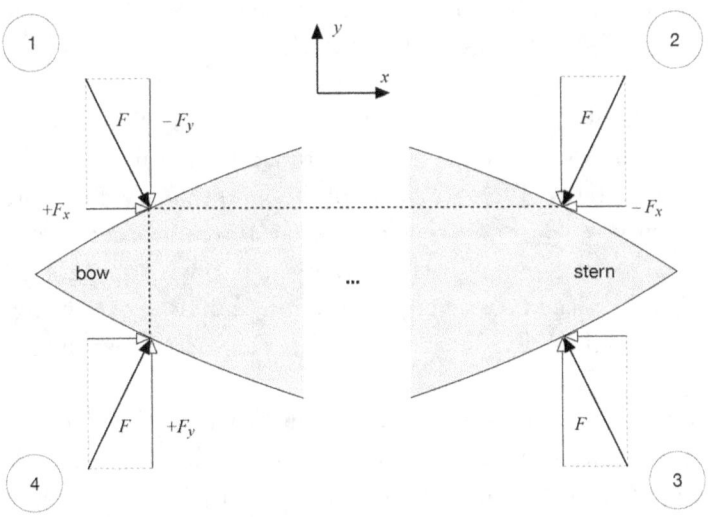

Figure 2.7: Pressure force vector components.

Drawing a line across the hull's width between locations 1 and 4 shows that the y-axis

---

12 As opposed to, say, an airplane wing, which generates lift owing to its asymmetric cross-section with lower pressure over a curved top surface (which sees higher speed flow) and higher pressure over a flat bottom surface (which sees lower speed flow than the top).

(transverse) components of force F_y have equal magnitude but opposite direction, and thus cancel. This will be true for any location along the length of the hull; try it for yourself and see. Then, drawing a line along the length of the hull between locations 1 and 2 shows that the x-axis (lengthwise) components of force F_x have equal magnitude but opposite direction, so these cancel as well. This result holds for any location along the hull's length; again, try it for yourself and see. It is also true for hulls with squared sterns. As a result, there is no net x- or y-axis force on this hull consistent with this simple model, e.g., no frictionless drag.

A simple Bernoulli analysis does not tell us anything about the drag force since it implies – within the assumptions of the model – that there is no net force on the hull to accelerate it (from propulsion) or decelerate it (from drag). So instead, we'll next consider a model that incorporates real-world phenomena. Like a wake, which is conspicuously missing from Fig. 2.6.

Of Wakes and Velocity Deficits

Those of us who have ridden very close to the stern of another paddled hull know that the flow behind the lead boat isn't smooth and uniform; it's nothing like the flow depicted in Fig. 2.6. There are whorls and eddies, and some inflow directed toward the lead boat's stern – this inflow is in part what we're riding when on the stern. Most of us would recognize that there is a wake behind the lead boat, and this wake gets wider and more diffuse the further back we are.

The stern wake behind a hull is shown conceptually in Fig. 2.8's plan view. As in our Bernoulli analysis, we adopt a paddler-centric frame of reference, which transforms hull speed U into a uniform fluid inflow of the same speed having parallel streamlines impinging on a stationary hull. The wake has a fan shape astern of the hull. Across the width and height of the wake, the flow velocity decreases compared to the freestream speed U. The wake also varies with distance from the stern, so the average velocity profile – the dependence on the average flow speed in the wake as a function of position – shown in the figure is at a fixed distance astern. The particulars of the fluid velocity distribution there aren't important for our purpose. Just note that, on average as we move across the wake, the flow velocity u is *less than* the freestream speed U. This is in part what appears to pull us forward if we are riding near the lead boat's stern: The trailing hull is experiencing "slower" water than the lead boat. And finally, outside the wake, we observe that the flow has uniform freestream speed U.

Having laid out the particulars, we can determine what role the wake plays in drag. We do this using the principle of mass conservation, along with Newton's 2nd and 3rd Laws. Newton's 3rd Law requires that any net force on the hull – here, the drag force – must be equal to the fluid mechanical forces acting on it. These forces arise from Newton's 2nd Law, which in its original wording states

> The rate of change of momentum of a body is equal to the resultant force acting on the body, and takes place in the direction of the force.

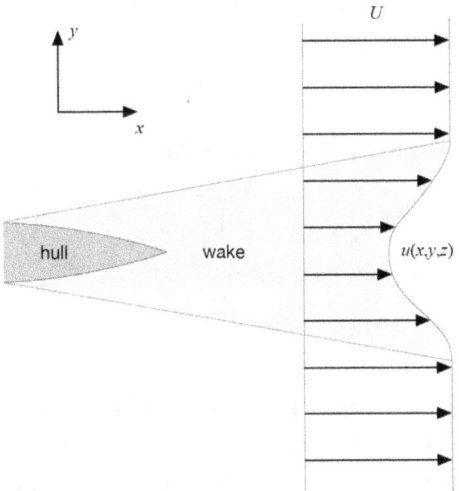

Figure 2.8: Plan view of the wake behind a hull.

Here, the "body" is the mass of fluid that passes around the hull. The hull's presence changes the flow's momentum since the wake's flow speed u is less than the free-stream flow speed U. This momentum difference leads to a net force on the hull,

$$F = \int_{A_2} \rho u(U - u)\, dA \,, \tag{2.3}$$

The integral sign '$\int$' indicates that the terms to its right is merely a sum over the wake's cross-sectional area A_2. Here, ρ is the water density.

Let's look at some limiting cases to see if this result makes sense. First, if the wake velocity u is everywhere equal to the freestream fluid velocity U then their difference $U-u$ equals zero, and the form drag also goes to zero. This limiting case is our wake-free Bernoulli model, which resulted in no drag force. Check!

Since the wake fans out from the stern, the average wake velocity over its cross-sectional area must *always* be less than the free-stream velocity U. Think of the wake as the widening part of a funnel: Higher-speed water enters upstream over a smaller cross-sectional area, and lower-speed water exits at the wider end because the water's mass is conserved. Consequently, the drag force is always greater than zero, *even in the absence of friction*. The drag force derived above is the form drag; the term $U-u$ is called the *wake deficit* or the *velocity deficit*.

Form drag arises because objects like kayaks and SUPs generate a wake, causing the flow's momentum to change. And because of Newton's 2nd Law, changes in momentum give rise to forces. The form drag's particulars *depend on the hull design, and are specific to every different hull*, just like the form drag of your hand changed depending on its shape (e.g., orientation) as you moved it through the air, or how your hand has a different size and shape than someone else's.

As we've seen, our tandem paddler's liquid goo doesn't affect friction drag, and cannot affect form drag because it doesn't change the hull shape. Plus, form drag is an inviscid effect. Best to look elsewhere for drag reduction.

But Wait - There's More: Novel Drag Reduction Techniques

In the 1980s several studies characterized exotic approaches to turbulent drag reduction, such as compliant skins, injecting long-chain polymers or air bubbles into the flow, and "scoring" surfaces along the direction of flow. In summary:

- The jury is still out on whether compliant (e.g., elastomeric) hull surfaces reduce drag. But it was a novel technology to explore and funded two years of graduate school for me.

- Injecting a cloud of tiny air bubbles into a boundary layer can considerably decrease drag. This dense layer of bubbles encases the hull in a thin envelope of air, so in effect the drag force comes primarily from moving the hull through air rather than through water. This entails less friction. Rumor has it that this technology has been incorporated into high-speed torpedoes. Incorporating an air injection system into our hulls would entail a lot of equipment, weight, and space, so it's not very practical. The displacement penalty from the added weight would more than offset any reduction in friction drag, as we saw in Chapter 1.

- Injecting long-chain polymers into a turbulent boundary layer can measurably reduce drag. It works by dampening the growth of turbulence in the cross-stream direction. And by "long-chain polymers," think of something like cooking oil. For this to work, we would have to controllably inject thin streams of it into the boundary layer the entire time we are underway. I doubt I'll be seeing pit stops at the General Clinton Canoe Regatta or the Ausable Marathon with crews re-supplying water, food, and canola oil.

- 3M sold an adhesive-backed polymer sheeting that incorporated parallel, triangular grooves ("riblets") that look like a corrugated roof in miniature and mimic certain fast-swimming sharks' skin. The grooves were 6 thousandths of an inch tall, spaced 6 thousandths of an inch apart. The sheeting was adhered to hulls with the grooves aligned in the direction of flow. Under certain conditions, surfaces with this ribbed covering experienced a drop in friction drag ranging from 2 to 8 percent. It was tested on rowing shells. Unfortunately, while there seemed to be some initial effect in reducing drag, the effect disappeared the next day. This change may be attributable to surface fouling between uses. A 6 thousandth of an inch feature is small, easily damaged, and can fill with river slime. Once that happens, friction drag can increase compared to an uncoated hull. I have heard of more lasting results from the '80s applying the 3M sheeting to sprint kayaks but have yet to find any published article or quantitative test data.

- NASA received a patent (U.S. #4,706,910) for a combined riblet and LEBU (Large Eddy Break-up Unit) system. LEBU's look like tiny spoilers, similar to the ones on Formula 1 race cars but much smaller. LEBUs control the dynamics of the outer turbulent boundary layer to reduce drag and influence the inner boundary layer in the process. While these might be useful on the back half of a hull, they run the risk of fouling, as well as damage from close encounters with logs and rocks.

Take-Aways

- The total drag force acting on our hulls is the sum of friction, form, and wave-making drag. Drag always opposes our motion.
- The no-slip condition requires that water molecules in contact with a moving hull "stick" and have zero relative velocity there.
- The water's speed varies with distance from the hull in a thin enveloping region called a boundary layer. Most of the boundary layer over our hulls is turbulent. This turbulence is the main contributor to friction drag.
- Friction drag is proportional to the hull's wetted surface area. A smoother hull will have a slightly lower frictional drag. However, we should consider any decrease in friction drag from a smoother hull in light of the contributions from form and wave-making drag.
- Form drag is a consequence of Newton's 2nd Law of Motion: when water flows past our hulls, the hull changes the flow's momentum and creates a wake. This change in the flow's momentum gives rise to a force, which is form drag. We tend to think of the 2nd Law as forces changing a body's momentum. Turns out it works in the other direction, too.

Extra Credit: Wake Deficit Derivation

Consider the hull and water system depicted in Fig. 2.9. A box-shaped slug of water called a *control volume* surrounds the hull below the waterline, with various labeled areas along the box's surfaces. The water volume's particular shape isn't important; a rectangular shape was chosen merely for convenience.

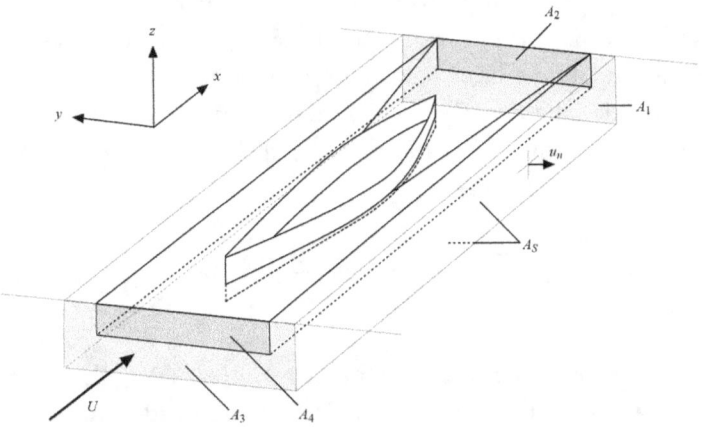

Figure 2.9: Hull perspective view with control volumes.

Downstream of the hull, a fan-shaped wake subtends an area on the control volume surface denoted as A_2. This area lies far enough downstream of the hull such that the flow speed outside of it is equal to the inflow speed U. There is also a U-shaped area A_1 that covers the balance of the control volume's downstream face that is coplanar with but nowhere overlapping A_2. Upstream of the hull is an area A_4 that is a collinear, equal-area projection of the downstream area A_2. A U-shaped area A_3 covers the balance of the control volume's upstream face that is coplanar with but nowhere overlapping A_4. And finally, the other sides of the box have area A_S. In total, these surfaces define the boundaries of the control volume. We are interested in the balance of forces that arise in this volume due to the flow of water in and out.

While water may enter and exit the control volume, it is neither created inside this space, nor does it somehow disappear there, either – there are no sources or sinks. Consequently, any mass flow of water into the control volume must be balanced by an equal amount of mass flow out of it. This balance reflects the fluid mechanical principle of *continuity* – what comes in must come out, somewhere, and needs to be accounted. Continuity is fluid mechanics' version of mass conservation.

The mass of fluid flow, sometimes called the *mass flow rate*, is defined as the product of the water's density ρ and its fluid velocity, summed over the areas through which it passes. Consequently, the mass flow rate has units of mass divided by time. Fig. 2.9 shows that water can flow through the upstream areas A_3 and A_4 with inflow velocity U, and consistent with our assumption of a uniform inflow, it flows out of the area A_1 with velocity U as well. Water flows out through the wake's cross-sectional area A_2 at a different velocity u, as indicated in Fig. 2.8. And finally, any fluid "leaks" from the control volume anywhere else do so with flow velocity u_n, directed perpendicular to any of the top, bottom, and side surfaces of the volume.

This summing over an area is another way of saying *integration*. Integration is a mathematical operation that adds together data over some portion of the data's range. Here the ranges are the various areas through which the fluid flows, and the data is mass flow rate at every infinitesimal point within those areas. The symbol for integration looks like an elongated and italicized 'S.' The integral's subscript specifies the range over which integration is performed; its argument defines what is being integrated.

For the five areas defined over the control volume, mass conservation requires that the two inflow terms (the positive ones) and three outflow terms (the negative ones) balance. This is expressed mathematically as

$$\int_{A_3} \rho U \, dA + \int_{A_4} \rho U \, dA - \int_{A_1} \rho U \, dA - \int_{A_2} \rho u \, dA - \int_{A_s} \rho u_n \, dA = 0 \cdot \qquad (2.4)$$

The differential dA denotes each of the infinitesimal areas over which integration is performed. Since the areas A_3 and A_1 are equal, the first and third terms cancel, leading to an expression for conservation of mass flow for our system as

$$\int_{A_2} \rho(U-u)dA = \int_{A_s} \rho u_n \, dA \, . \tag{2.5}$$

Next, Newton's 3rd Law requires that any net force on the hull – here, the drag force – must be equal to the fluid mechanical forces acting on it. There are no other forces in our model. The fluid mechanical forces arise from Newton's 2nd Law, which states

> The rate of change of momentum of a body is equal to the resultant force acting on the body, and takes place in the direction of the force.

Here, the "body" is the mass of fluid that passes through one of the surfaces defined above. Its momentum P over a differential area dA is the product of its mass and velocity,

$$P = \rho u dA \, . \tag{2.6}$$

The rate at which momentum enters or leaves one of the surfaces is the product of momentum times velocity, which over an infinitesimal area dA equals

$$\rho u^2 dA \, . \tag{2.7}$$

Summing the components of the rates of momentum change over their respective areas gives us an expression for a drag force F – the only force in the control volume aside from those arising from the dynamic pressure of the fluid – in terms of the forces over each of the control volume's five bounding areas,

$$F = -\int_{A_3} \rho U^2 \, dA - \int_{A_4} \rho U^2 \, dA + \int_{A_1} \rho U^2 \, dA + \int_{A_2} \rho u^2 \, dA + \int_{A_s} \rho U u_n \, dA \, . \tag{2.8}$$

The free-stream velocity U is constant over the y-z plane outside of the wake. Consequently, we can move U outside of the integrals. Canceling terms and substituting from the expression we derived for fluid mass continuity yields a simplified expression for the form drag force:

$$F = \int_{A_2} \rho u(U-u)dA \, . \tag{2.9}$$

Since all other terms in the integrand are constant, the integral averages the velocity u and its square over the area A_2.
    Q.E.D.

Further Reading

Wen-Hsiung Li and Sau-hai Lam, *Principles of Fluid Mechanics*, 8th Edition, Addison-Wesley (1978).

Baljit Singh Sidhu, Mohd Rashdan Saad, Ku Zarina Ku Ahmad, and Azam Che Idris, "Riblets for Airfoil Drag Reduction in Subsonic Flow," *ARPN Journal of Engineering and Applied Sciences* **11**(2) (2016).

D. M. Ladd, J. J. Rohr, L.W. Reidy and E.W. Hendricks, "The Effect of Riblets on Laminar to Turbulent Transition," *Experiments in Fluids* **14**, pp. 109 (1993).

H.L. Petrie, S. Deutsch, T.A. Brungart, A.A. Fontaine, "Polymer Drag Reduction with Surface Roughness in Flat-Plate Turbulent Boundary Layer Flow," *Experiments in Fluids*, **35**, pp. 8-23 (2003).

Zaiguo Fu and Yasuo Kawaguchi, "A Short Review on Drag-Reduced Turbulent Flow of Inhomogeneous Polymer Solutions," *Advances in Mechanical Engineering*, **2013**, Article 432949 (2013).

P. Henrik Alfredsson and Ramis Örlü, "Large-Eddy BreakUp Devices – a 40 Years Perspective," *Flow, Turbulence and Combustion*, **100**, pp. 877-888 (2018).

D. W. Bechert, M. Bruse, W. Hage, J. G. T. Van der Hoven, "Experiments on drag-reducing surfaces and their optimization with an adjustable geometry", *Journal of Fluid Mechanics* **338**, pp. 59-87 (1997).

Oliver Rohr and Franz Gammel, "Method for Producing a Surface of a Component, Said Surface Having a Reduced Drag and Component with Reduced Drag," U.S. Patent Application No. 22014/0130318 A1 (May 15th, 2014).

Dennis Bushnell and Jerry Hefner, Eds., *Viscous Drag Reduction in Boundary Layers*, American Institute of Aeronautics and Astronautics, Washington D.C. (1990).

CHAPTER 3

Wave Drag and Shallow Water

Introduction

We all know that paddling in shallow water is a pain. You trim aggressively bow down to keep from experiencing that "sinking feeling" in the stern, and paddle like there's no tomorrow. Yet when you hit the shallows, it still feels like you've hit a wall. Waves become steeper; your hull speed falls. And heaven forbid if you're on someone's inside as you enter a shallow water turn. Hello, shoreline!

What is it that makes shallow water so challenging to paddle? Why does your hull slow down? Why do waves break toward shallow water and then steepen? More fundamentally, why are the physics governing hulls in shallow water any different than in deep water? After all, it's still just water – isn't it?

Yes, it's all water. But we inhabit a very particular environment when we paddle: the water's surface. It is because of this interface between air and water that our challenges arise. The culprit is waves.

We see waves all around us when we're on or around water: the expanding wave patterns created by a pebble cast into the water; our slow rise and fall in the long-wavelength swells created by a powerboat on a lake's far shore; the respite offered by surfing the waves behind another boat during a race. And what dominates paddling resistance at higher speeds are the surface waves caused by varying pressures around our hull as it moves through the water. These waves' character changes when the depth of water beneath our hulls is less than about half the boat's length, which further increases wave resistance.

Of Waves and Wavelengths

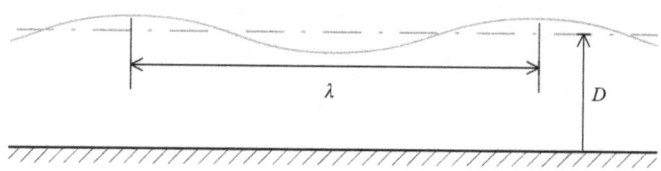

Figure 3.1: Water and wave geometry.

Consider waves moving over the surface of water shown in Fig. 3.1. The water has average depth D. The distance between two wave crests is the wavelength, represented by the Greek symbol λ ("lambda"). We know from experience that waves such as this don't just sit there, but travel at a certain speed which we represent by c_p. There is a relationship between this *phase speed* c_p, the wavelength λ, and the depth D,

$$c_p = \sqrt{\frac{g\lambda}{2\pi} \tanh\left(\frac{2\pi D}{\lambda}\right)}. \tag{3.1}$$

In this equation, g is the gravitational acceleration constant, π is the constant "pi" (which equals 3.1415...), and tanh is the hyperbolic tangent function. You don't need to worry about the hyperbolic tangent function. We'll be looking at deep and shallow water cases where this function reduces to something straightforward.

As you can see in equation (3.1), the water wave's speed is a function of water depth and the wavelength – plus a few constants – and nothing else. In other words, *wave motion is a property of the water having a free surface, whether there is a boat there or not.* And wave motion must satisfy this relationship: for a given wave speed, there is a corresponding wavelength. The only thing a boat does is provide what mathematicians call a *forcing function* at the water's surface and thus *initiate* waves.

When the water is deep, e.g., the depth D is much greater than the wavelength λ, the hyperbolic tangent function reduces to 1. Equation (3.1) is then greatly simplified as

$$c_p = \sqrt{\frac{g\lambda}{2\pi}}. \tag{3.2}$$

Equation (3.2) shows that in deep water, longer-wavelength waves move faster than short-wavelength waves. This is familiar to anyone who has sat on the shore of a lake: when powerboat waves approach, longer swells arrive first. Also, equation (3.1) shows that deep-water waves do not depend on the depth D, but only upon the wavelength λ. For the next bit, we'll use this *deep-water approximation* to guide our analysis.

We've all seen the wave train spreading out beside and behind a hull moving across deep water, as depicted in Fig. 3.2. These so-called *Kelvin-Froude waves* comprise divergent waves spreading out from the hull and transverse waves that span the divergent wave fan.

The wave pattern depends on boat speed. Waves created by the hull must satisfy equation (3.2), which relates wavelength λ and the wave's phase speed c_p. As the hull speed increases, the waves created by the hull follow the hull at that same speed. And those waves, which are a property of the water, must satisfy equation (3.2). Consequently, the wavelength will increase as hull speed increases.

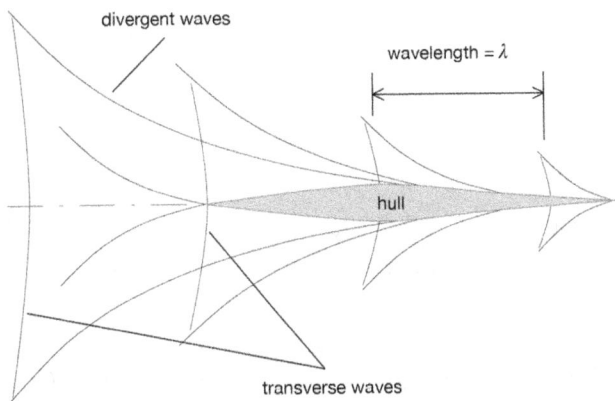

Figure 3.2: Kelvin–Froude waves.

Wave-making due to hull motion leads to wave patterns alongside the hull, as depicted in Fig. 3.3. Fig. 3.3 shows a profile view of an 18'6" (5.64 *m*)-long hull, paddled over a range of speeds such that the transverse bow wave's wavelength changes in relation to the hull's length. In Fig. 3.3(a) the bow wavelength is half the hull length; this corresponds to a paddling speed for the hull of about 4.7 *mph* (7.56 *kph*). As the speed increases to about 5.4 *mph* (8.7 *kph*) the wavelength is now about two-thirds as long as the hull. Finally, when the paddling speed reaches about 6.7 *mph* (10.78 *kph*), the wavelength equals the hull length.

Throughout, the transverse bow wave and stern wave interact. Because of *superposition* these waves combine constructively and destructively, leading to the wave pattern shown in Fig. 3.2. As speed further increases, a trough forms in the wave pattern just aft of midships. As a result, the stern begins to sink since there is less water to support the hull's rear half buoyantly. This sinking is why you need to trim a hull slightly bow down even in deep water; when the stern sinks, the hull becomes less hydrodynamically efficient.[13] Finally, as the hull speed surpasses 7 mph, as shown in Fig. 3.3(d), the superposed wave pattern causes the stern to sink further, requiring significant bow down trim to maintain hydrodynamic efficiency – plus a lot of hard paddling to maintain this sprint speed!

The speed at which the water wavelength equals the hull length is called the *hull speed*. The hull speed, described by Froude in 1868, is a rule of thumb describing the approximate

---

13 Some designers move the hull's widest point aft of midships to address this shift in buoyancy, as well as other hydrodynamic effects.

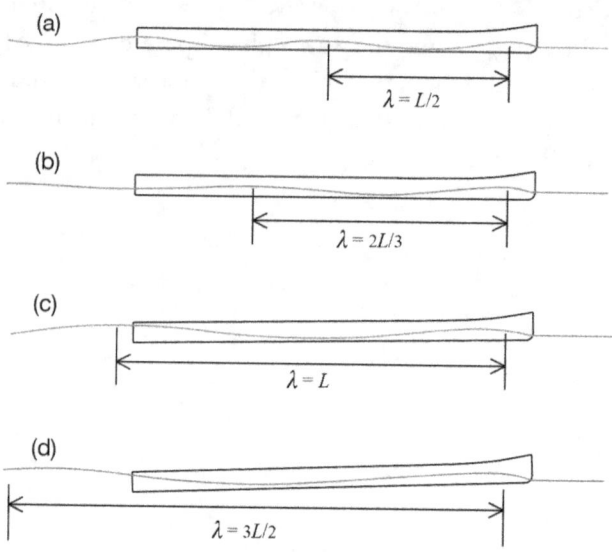

Figure 3.3: Hull in its own wave train at increasing speeds (not to scale).

maximum efficient speed for a displacement hull.[14] Since most canoes and kayaks are displacement hulls, this hull speed is not a speed limit but rather an indication of when wave drag begins to dominate total drag. For example, a slender 18'6" racing hull can be paddled beyond 6.7 *mph* (10.78 *kph*); it just takes a fair amount of fitness to maintain that speed. Note that the hull speed varies as the square root of the hull's waterline length; we find this using Equation 3.2 when λ = *L*. Consequently, all other factors being equal, a longer hull tends to be a faster hull.

We can characterize this wave drag effect using a non-dimensional ratio called the *Froude Number*. The Froude Number F_r is the ratio of hull speed *V* to the phase speed c_p (which, in the deep-water case, corresponds to λ = *L*):

$$F_r = \frac{V}{c_p}.$$ (3.3)

As a rule of thumb, a higher Froude Number entails greater wave resistance because of the amount of energy transferred into the divergent surface waves. The case depicted in Fig. 3.3(c) corresponds to $F_r = 1$. When $F_r > 1$, you're paddling faster than the phase speed, e.g., faster than the water "wants" to go per equation (3.2), which will significantly increase wave drag's contribution to the total drag force.

Waves and Conservation of Energy

So why does a hull moving through water cause waves in the first place? If we could some-how prevent wave creation, would we escape the effects of wave drag? Unfortunately, like

---

14 A displacement hull displaces water; a planing hull planes over the water's surface when up to speed.

death and taxes, you can't get around the laws of physics. Waves arise because the shape of the hull – *any* hull – deflects water. To demonstrate this, consider a hull moving at speed V_0, as shown in Fig. 3.4.

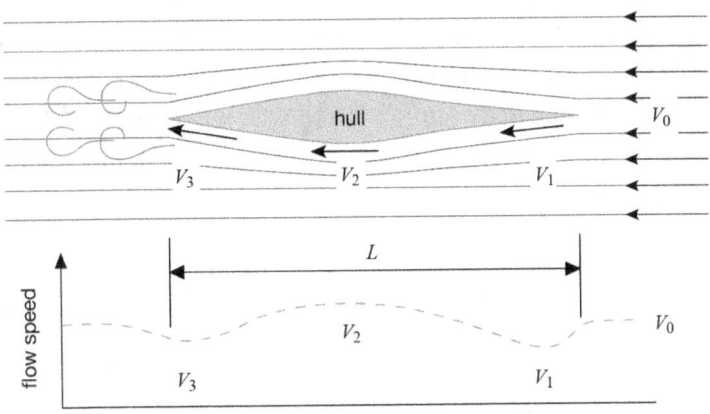

Figure 3.4: Flow speeds adjacent to hull (not to scale) and corresponding flow speed plot.

Once again, we adopt a paddler-centric reference frame. In this reference frame, the water appears to approach the hull at speed V_0. The water that impinges upon the side of the canoe near the bow slows down a bit to speed V_1. As the water continues to flow past the hull, it accelerates a bit around the location of maximum waterline width, reaching a speed v_2 that is a bit higher than your paddling speed V_0. And finally, as the water approaches the stern, it slows to speed V_3, which is a bit slower than the speed adjacent to the bow because some of the water's energy is spent creating a wake. This variation in flow speed along the hull is depicted conceptually by the dashed line plotted in the lower half of Fig. 3.4.

The water impinging on the bow has a certain amount of *kinetic energy*, e.g., energy due to motion. The kinetic energy of a unit mass of incoming water is expressed by

$$\frac{mV^2}{2},\tag{3.4}$$

where m is the mass of water. Since energy is conserved, the kinetic energy at location 1 with water speed V_1 must equal this "incoming" energy. However, since V_1 is less than V_0, and energy is conserved, the decrease in kinetic energy at location 1 must be augmented by increasing *potential energy* there.

We can think of potential energy as stored energy. When you throw a ball into the air, the instant it leaves your hand the ball has a certain amount of kinetic energy. As it rises above you, the ball gradually slows down until its vertical speed equals zero. At this precise moment, the baseball has potential energy – only potential energy – which equals the amount of kinetic energy you imparted the moment the ball left your hand.[15] This equivalence is due to *conservation of energy*. As the ball falls to earth, its potential energy is converted back into kinetic energy, and the ball picks up speed. It will return to your hand at the same speed as when you initially threw it; when it lands, it will have the same amount of kinetic energy as when you threw it upwards because energy is conserved.

Like the ball, water will have potential energy if you raise it up. The total energy at location 1 in Fig. 3.4 is the sum of the local kinetic and potential energy, which must equal the initial amount of kinetic energy of the incoming flow having speed V_0:

$$\frac{mV_1^2}{2} + mgh_1 = \frac{mV_0^2}{2} . \tag{3.5}$$

The height h_1 is the displacement of water at location 1 about the average water depth D as shown in Fig. 3.5.

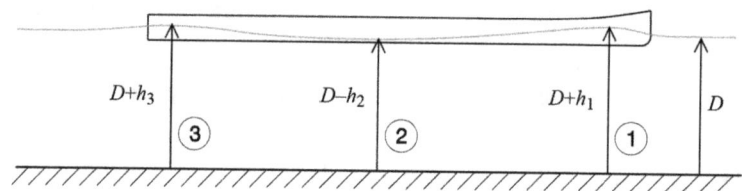

Figure 3.5: *Profile view of hull with hull speed wave train (not to scale).*

Conservation of energy requires that the water height h_1 be greater than the zero, e.g., the water is displaced above the average depth D. A similar result occurs at location 3. Since the local water speed V_3 is a bit less than V_1, the height h_3 is a bit lower than h_1. Conversely, at location 2 the local water speed V_2 is greater than the incoming water speed V_0. This local increase in the water's speed means that the kinetic energy at location 2 is greater than the incoming kinetic energy. Since energy is conserved, the potential energy at location 2 must offset this increase in kinetic energy so that

$$\frac{mV_2^2}{2} - mgh_2 = \frac{mV_0^2}{2} . \tag{3.6}$$

In this case, the height h_2 lies below the line of average depth D; the water height is lower to satisfy energy conservation.

What does this water profile along the hull in Fig. 3.5 look like? That's right: waves. Because your hull deflects and slows water, then accelerates it, then slows it again you create a disturbance in the water. And, since the water inherently "likes" to support and sustain surface waves, these local variations in water height caused by your hull become the forcing functions that drive wave creation.

---

15 Minus any energy lost due to air drag.

Designers of modern high-performance hulls have deduced many crafty ways of minimizing wave creation and wave drag. And these high-performance hulls allow you to paddle well beyond the theoretical hull speed. But the underlying principles still apply for all hulls: water is still water, your hull displaces it when underway, and the laws of physics hold no matter what. There will be waves.

Shallow Water

Now all this wave stuff is great, but when does water become *shallow*? Isn't that the point of this chapter? Indeed, it is; we just had to cover the preliminaries. A lot of preliminaries!

Return now to equation (3.1), which defined the phase speed of surface waves. Previously we had concerned ourselves with the deep-water limit, where the expression for the phase speed took a simple form. This simplification is valid when the hyperbolic tangent's argument equals about 3; you can check for yourself by plugging in a few numbers. Consequently, water is "deep" when

$$\frac{2\pi D}{\lambda} \approx 3 \rightarrow D \approx \frac{\lambda}{2}, \tag{3.7}$$

where we have recognized that π is roughly equal to 3. Equating the wavelength λ to the hull length L, equation (3.7) indicates that water is deep when it is deeper than about half the hull's length. "Deep" or "shallow" are always defined in reference to something. We see that deep or shallow are defined in terms of wavelength and/or hull length.

In the shallow water case, where the depth D now becomes much less than the hull length L, the expression for the surface wave phase speed (3.1) simplifies to

$$c_p = \sqrt{gD}. \tag{3.8}$$

In other words, in very shallow water the phase speed has nothing to do with the surface wave's wavelength; all waves move together, without dispersion. It only depends on the gravitational constant g and the depth D. As the shallow water depth decreases, the phase speed decreases as the square root of depth. Things have changed! But what happens in between? Is the transition from deep water behavior to shallow water behavior abrupt or smooth?

We can compute and plot the phase speed versus depth to answer these questions. We shall assume that the wavelength in equation (3.1) corresponds to the length of an 18'6" hull running at 6.65 *mph* (10.7 *kph*). The resulting data is plotted in Fig. 3.6. As you can see, when the depth equals the boat length, or even about half the length, the phase speed equals the hull speed – the plot stays flat. The phase speed begins to decrease when the depth is about half the hull's length – as expected in light of our analysis – and starts to drop significantly when the water depth is approximately one-third to one-quarter of the hull's length. This depth corresponds to what paddler's refer to as "concrete water," about 4' to 5' deep (1.22 – 1.52m). As the water depth decreases further, the phase speed drops precipitously.

A drop in phase speed indicates that water intrinsically supports more slowly moving waves as it becomes shallow. If you try to paddle faster than the phase speed, you provide a forcing function that tries to make the water move faster than it wants to, and you pay the price in wave resistance.

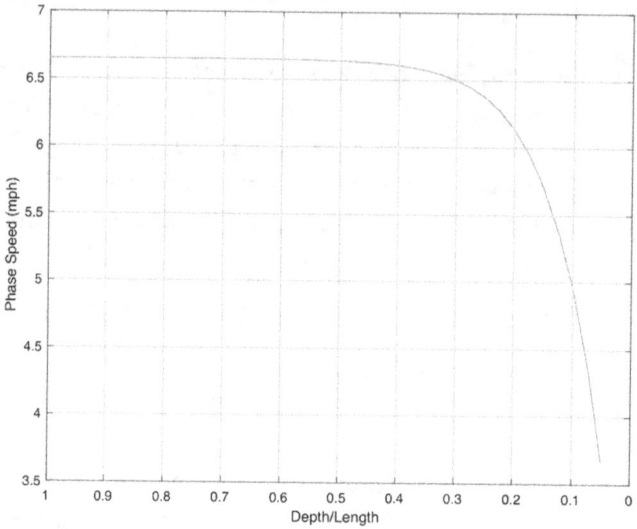

Figure 3.6: Phase speed vs. depth for an 18'6" hull.

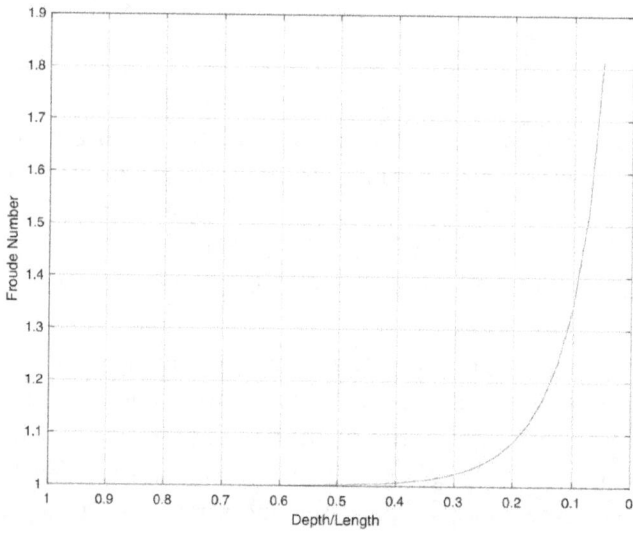

Figure 3.7: Froude Number vs. normalized depth.

We can appreciate this increased wave drag effect by plotting the Froude Number vs. depth, as shown in Fig. 3.7. Recall that the Froude Number expresses the ratio of hull speed to the phase speed and reflects increasing wave resistance. Again, we consider the case of an 18'6" long hull traveling at 6.65 *mph*. Fig. 3.7 indicates that when the water depth drops below 2' (0.61*m*), wave drag forces on the hull begin to skyrocket. Sound familiar?

Other factors come into play in shallow water besides decreasing surface wave phase speed. The water beneath the hull becomes "squeezed" between hull and bottom, which are now close to each other. This causes the water beneath and around the hull to accelerate. This acceleration leads to a suction force on the hull because of Bernoulli's Principle. The downward force acts as if the paddlers suddenly gaining weight, and the boat is pulled lower in the water. And as we learned in Chapter 1, a heavier team is a slower team, and now you are pushing a larger wetted surface area through the water than before you hit the shallows.

Further, in shallow water surface waves grow in height. With increasing wave height (and wave slope) nonlinear effects take over. Our analysis relies upon the so-called *small wave approximation*, which begins to break down when the wave height is no longer small compared to the water's depth. At least it was fun while it lasted.

Shallow Water Turns

And finally, about those shallow water inside turns. Water tends to grow shallow as you get closer to shore, especially on the inside of river turns. Shallow water effects become more pronounced closer to shore. We have seen in Fig. 3.6 that the phase speed decreases as water becomes shallower. One can construct a "map" of wave patterns over a river's width using this information, as suggested in Fig. 3.8.

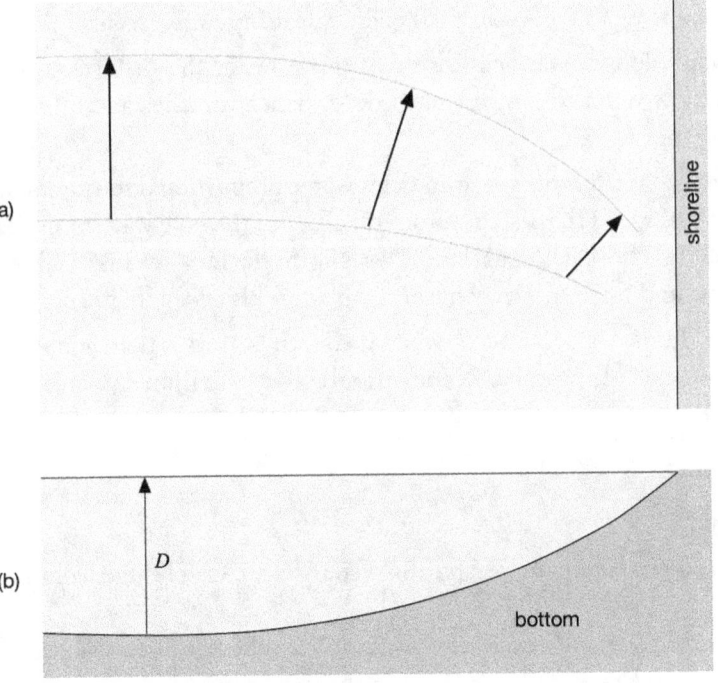

Figure 3.8: Wave refraction, (a) top view, and (b) water/bottom cross-section.

In Fig. 3.8(a) the arrows represent the phase speed at the depth corresponding to their location. A longer arrow represents a greater phase speed; they also span one surface wave cycle's wavelength. As the water becomes shallower, the phase speed decreases, as does the wavelength. The wave velocity turns toward the shore since there must not be any discontinuities in the wave pattern. Thus, the surface waves themselves *turn toward the shore* as the bottom becomes shallower there.

This phenomenon is called *refraction*, analogous to the bending of light waves at an air/water interface. Viewing a pencil in a glass half full of water is a familiar example of this wave bending phenomena, albeit with light waves rather than water waves. So, as the surface water waves bend, they want to take you with them: toward the shore. And as if refraction wasn't enough, because the wavelength shortens, the wave heights must increase to satisfy conservation of energy. This is why you want to lead the way into a shallow inside turn rather than follow on a leading boat's inside shoulder. Surf's up!

Take-Aways

- Water supports waves along its surface, and the character of these waves changes with depth.

- Hulls moving over water create waves because of energy conservation. From the hull's frame of reference, the inflow of water slows at the bow, creating a bow wave. The bow wave reflects a local increase in the flow's potential energy since the water is raised about the waterline there. Faster moving water downstream of the bow results in a wave trough. Faster water indicates a local increase in the water's kinetic energy. Increased kinetic energy requires a decrease in local potential energy, leading to the trough alongside our hulls.

- In deep water, longer-wavelength waves move faster than shorter-wavelength waves. In very shallow water this wavelength dependence vanishes and all wavelengths move at the same speed.

- Hulls have a characteristic speed in deep water proportional to the square root of their length. At this speed the distance from the front of the bow wave to the back of the wave trough alongside the stern equal the hull's length; the hull just fits inside one wavelength. Very narrow and fine hulls may naturally move a bit faster than this.

- For paddled hulls, water is "deep" when its depth is greater than about half of its length at the waterline. As water becomes progressively shallower, wave resistance grows nonlinearly.

Further Reading

James Lighthill, *Waves in Fluids*, Chapter 3: Water Waves, Cambridge University Press (1980).

G.B. Whitham, *Linear and Nonlinear Waves*, John Wiley & Sons, New York (1974).

CHAPTER 4

What Moves You

Introduction

When you paddle, each stroke propels your hull forward against our familiar nemesis, drag. But why does the hull move forward and not in some other direction? And more fundamentally, why does it move at all? The principle of momentum provides the answer. And space boots.

When I was in 6th grade, I bought a book of science fiction stories at a garage sale. One of the tales centered on an astronaut marooned outside his spacecraft; how he became marooned, I don't recall. Fortunately, he was wearing a spacesuit – it's pretty cold in outer space. Unfortunately, the spacesuit had no thrusters to propel him back to his craft. And he had a limited supply of air left to breathe. How did he get back inside his rocket ship before running out of air? By using science!

Our stranded astronaut recalled that when you fire a rifle, there is recoil. The rifle flings a bullet in one direction, and in response, the rifle moves in the opposite direction. The astronaut realized that if he threw something in the direction opposite his spacecraft, his body would recoil in response and move toward the ship. Now he couldn't throw his oxygen tanks (gotta breathe), his space helmet (to avoid the worst case of brain freeze ever), or his space gloves (he needed functioning hands to open his spacecraft's hatch upon arrival). So, he took off his boots and threw them in a direction away from his rocket ship as hard as he could. He had a pretty good aim because he drifted slowly toward the ship's hatch and was able to climb back inside. All because he remembered physics and the principle of *momentum conservation*.

On Momentum

We've all heard the word momentum, but what does it mean in scientific terms, and how does it apply to paddling? Momentum can be thought of as mass in motion. A hull and its paddler(s) have mass, so if the hull is moving, it has momentum. An object's momentum depends upon its mass and its velocity. Formally, momentum – typically written as P – is equal to the product of the object's mass and its velocity:

$$\text{Momentum} \equiv P = \text{mass} \times \text{velocity} . \tag{4.1}$$

Momentum is linearly proportional to an object's mass, and linearly proportional to its velocity. And by linearly, we mean that neither quantity appears as a square, cube, square root, etc., but only by itself. This means that if you throw a baseball at 45*mph* (72.4 *kph*), and a major league pitcher throws the same baseball at 90*mph* (144.8 *kph*, e.g., twice as fast), then the baseball thrown by the major leaguer has twice as much momentum as yours.

So now consider a hull paddled by a solo paddler. We'll adopt a hull-centric coordinate system. At the end of each stroke, their hull has a certain amount of added momentum P_{hull}, which is quantified using the total mass of the hull m_{hull} – including the mass of paddler – times the change in the hull's velocity v_{hull}:[16]

$$P = m_{hull} \times v_{hull} . \tag{4.2}$$

With each stroke, the paddler imparts an increment of momentum to the hull above its current state. This is because the paddle blade accelerates and "throws" a volume of water. In an inertial, Earth-reference coordinate system your paddle appears to be planted in concrete at the catch, albeit with a little slippage. Yet if you watch the strokes of adjacent paddlers while underway, you'll note little pools of swirling, aft-moving water behind their blades following their power phase. Each stroke changes the momentum of the water entrained by the paddle, in the direction opposite the hull's motion.

Figs. 4.1-4.5 show side views of the power phase via a sequence of five photographs. The paddle rotates during this phase. It stays well planted while the blade is near the vertical, slipping slightly sternward in the inertial (Earth-based) reference frame. More importantly, in the hull-centric frame of reference, the blade moves a significant distance along the hull – note the number of white fiducial lines the paddle traverses during the stroke compared to the number in the inertial reference frame of the water. From the paddler's point of view, the blade has displaced a mass of water sternwards.

---

16 We are concerned here with changes in momentum with each stroke.

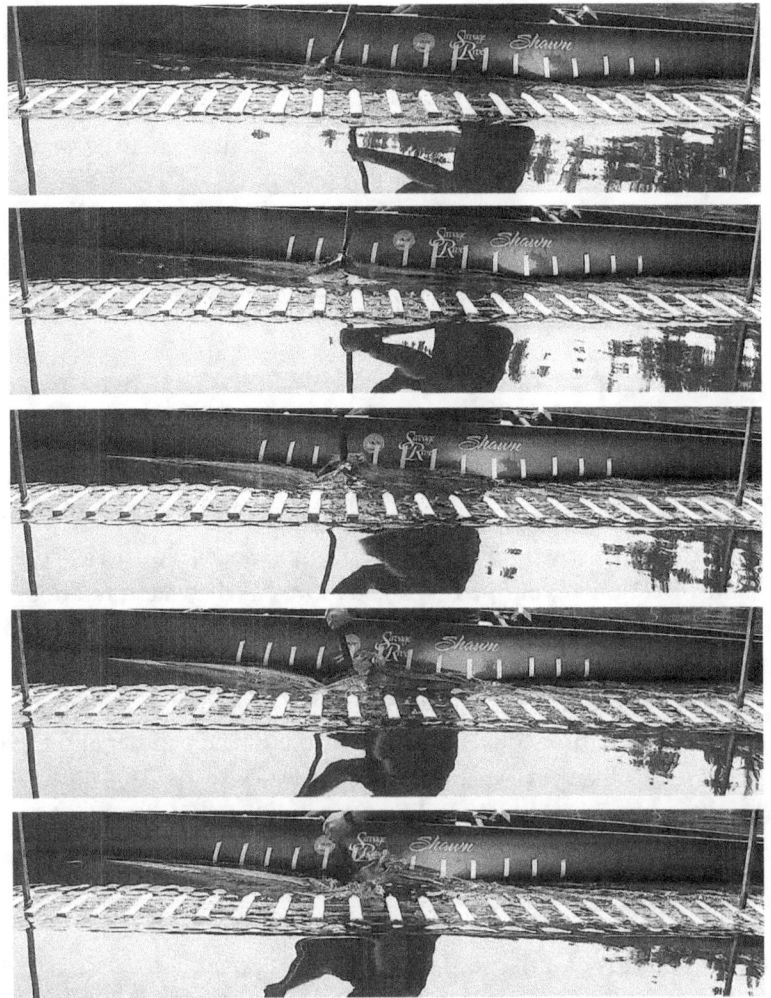

Figures 4.1-4.5: Power phase, side view with hull and fixed fiducials.

Your paddle stroke imparts momentum P_{water} to an entrained mass of water m_{water}, which we idealize as a plug of water having a velocity v_{water}, in the direction opposite your hull's motion.[17] The momentum of this idealized plug of water is consequently

$$P_{water} = -m_{water}v_{water}.$$ (4.3)

The minus sign indicates that the momentum imparted to water is in the opposite direction of the hull's momentum. In physics, velocity is a vector quantity: velocity has both a *magnitude* (referred to commonly as *speed*) and a *direction*. The minus sign here indicates that the displaced water and the hull move in opposite directions.

17 This is a simplification since the paddle imparts momentum – both linear and angular – to a mass of water it entrains. This mass is idealized as a chunk of water behind the paddle blade's power face. Reality is a bit more complex, as momentum is imparted ot the water in 3D. Ever notice nice paired whorls of water behind your paddle placements? Those are momentum embodied: mass in motion.

Physics teaches us that the momentum of a system is conserved. Here, the system is the hull and paddler, plus the mass of water "thrown" by the paddle. Consequently, at the conclusion of each stroke, the system's total momentum – here, the hull/paddler(s) and the paddled water – must equal zero. The total system momentum may be written as

$$P_{hull} + P_{water} = m_{hull}v_{hull} - m_{water}v_{water} \,.$$

(4.4)

As a result, the hull plus paddler's momentum, plus the momentum of the entrained water, must be equal, which means

$$m_{hull}v_{hull} = m_{water}v_{water} \,.$$

(4.5)

So, to move the hull, you impart momentum to the water entrained by your paddle. And the hull will travel in the direction opposite the water that your paddle has "thrown." This should be familiar to anyone who has paddled a canoe, kayak, or SUP. The hull doesn't move in the same direction as your paddle blade moves in the water, nor does it move perpendicular to the path of the paddle. Momentum has both a magnitude and a direction, and the system's total momentum is conserved.

Perhaps the most interesting takeaway, particularly relevant to our discussion, is that the paddled water's momentum equals the momentum of the paddler and their hull at the end of each stroke. It's like our astronaut floating freely in space – to move toward his spacecraft, the astronaut threw his space boots in the direction opposite his ship, and off he zoomed to safety. If he had heavier space boots or threw his boots faster, he could move faster in response.

So, what does this tell us about moving faster? Equation (4.5) shows us how. We can write equation (5) in terms of hull speed v_{hull} as

$$v_{hull} = \frac{m_{water}v_{water}}{m_{hull}} \,.$$

(4.6)

To increase hull speed, equation (4.6) gives us several options:

• For a fixed paddle size and power phase stroke speed, the numerator in equation (4.6) is a constant. So, to go faster, our paddler should get a lighter hull, lose weight, or both. The only remaining variable in this scenario is the combined mass of the hull and paddler.

• For a fixed power phase blade speed, our paddler could move a bigger slug of water by using a paddle with a bigger blade in order to entrain more water with the stroke and thus impart more propulsive momentum.

• For a fixed paddle size, our paddler could move the entrained water faster with a faster power phase blade speed or use a smaller paddle but have an even faster power phase speed.

• Or our paddler could lose weight, buy a lighter hull, get a paddle with a bigger blade, and move that blade faster. Phew! I'm tired already, not to mention broke from the cost of all the new gear.

Practically speaking, each option will appeal to specific paddlers based upon energy, paddle force, and power considerations. We'll address energy issues below; paddle blade force and power will be addressed in Chapters 6 and 7.

Now what if a canoeist isn't paddling, but instead is poling? As you may (or may not) know, canoe polers "stand tall in their canoes" and propel themselves using a pole of wood or aluminum, driving the pole into the riverbed, and pushing off as illustrated by twelve-time ACA poling National Champion Harry Rock in Fig. 4.6.

Fig. 4.6: Poling up a drop.

Assuming the pole is well planted and doesn't slip, the poler is pushing against the mass of the Earth – a pretty large mass. For the poler, equation (4.4) takes a particularly interesting form:

$$P_{hull} + P_{Earth} = m_{hull}v_{hull} - m_{Earth}v_{Earth} = 0 \, . \tag{4.7}$$

So, each time the poler plants their pole and propels their canoe forward, they are changing the Earth's momentum, pushing it in the opposite direction!

As noted above, a hull can attain a particular speed by using the paddle to move a larger mass of water slowly or a smaller mass of water more quickly. The choice is yours as long as total momentum is conserved. The difference between these two scenarios becomes more apparent when we consider the energy expended in each case, e.g., the cost in energy to *you*, the paddler.

At the end of each stroke, the hull/paddler/water system's energy is comprised of the motion of the hull and paddler(s) and the motion of the paddled water. This motion energy is the system's *kinetic* energy. Kinetic energy is linearly proportional to mass and proportional to the square of velocity.[18]   Assuming the hull is brought to the same speed using either paddle, the hull's kinetic energy is the same for both cases. A little analysis shows that the ratio of kinetic energy expended, which here equals the amount of energy imparted to the water using either paddle, is given by the ratio

$$\frac{E_{water,1}}{E_{water,2}} = \frac{A_2}{A_1}.$$
(4.8)

Since the larger blade's area A_2 is always bigger than the smaller blade's area A_1, the ratio of A_2 to A_1 will always be greater than 1. Equation (4.8) shows that you will *always* expend more energy paddling with a smaller blade than paddling with a larger blade to attain the same hull speed. The cost of using a smaller paddle to achieve the same hull speed is greater energy expenditure – by you. Surprise!

In both scenarios, the hull and paddler's mass and speed at the end of the stroke stays the same, hence their contribution to the system's total kinetic energy is fixed. But while the mass of water moved by the smaller paddles is less, the water's kinetic energy is only linearly proportional to the mass reduction. This gain is more than offset by the squared dependence of kinetic energy on the velocity. To achieve a given hull speed, the smaller blade must move its smaller entrained water mass faster because of momentum conservation. What you appear to gain in moving a smaller mass of water is more than offset by having to move it with a faster power phase speed. So, if you prefer a smaller blade use it briskly, and have a well-trained aerobic/anaerobic engine. This area advantage will have practical limits, dictated by how much force and power the paddler can exert. Most will find more than a 10% blade area increase challenging without changing their stroke mechanics.

I'll be the first to admit that I was initially puzzled by these results; smaller blades are the trend among many local paddlers. But thinking back to my own experiences in 2008, I realized that I had lived these equations. That Summer I purchased a new bentshaft paddle with a wider blade; the bend angle and shaft length were the same as my other paddles. It took me a while to zero in on how to take advantage of the new paddle, since I seemed to be pulling harder but not going any faster, and getting more tired in the process. Things got better when I spent some time paddling tandem with a much more experienced partner. In doing so, I learned to accelerate the blade more smoothly and employ a more complete power phase. I stopped recovering before the wider blade's velocity had reached steady state. In other words, a slightly longer stroke than the "choppy" stroke I had been using. A few sessions later, and suddenly my speed improved, with a slightly lower cadence and a lower heart rate, e.g., less energy expenditure. Granted, I'm mostly made of slow-twitch muscle fiber. But I could grind at a faster speed, using less energy, once I sorted out the bigger blade.

And I owe it all to space boots.

---

18 Formally, since velocity is a vector, this should be expressed as a dot product. For simplicity we are only considering motion in one direction, so velocity here is a scalar. Bonus points if you caught this.

Take-Aways

- Momentum is mass in motion. It is expressed as the product of an object's mass and velocity.
- The momentum of a system is conserved. This means that the system's total momentum must sum to zero. For paddling the system comprises the hull, paddlers, paddles, gear, plus the water engaged in our propulsive efforts.
- A hull's momentum is equal in magnitude but opposite in direction to the momentum imparted through our paddles to the water.
- While we can idealize the momentum imparted to the water by our paddles as a chunk of water we "throw," the actual fluid mechanics are a bit more complex. Just keep in mind that we're imparting momentum to the water, and this momentum equals the momentum imparted to our hulls opposite to the direction of our paddling stroke. The "chunk" of water is just an idealization.

Extra Credit

We'll use conservation of energy to determine the energy cost of different paddle blade sizes. The total kinetic energy of the hull/paddler(s)/water system is the sum of the kinetic energy of the hull and paddler, and the kinetic energy of the paddled water:

$$E_{kinetic} = \frac{1}{2}m_{hull}v_{hull}^2 + \frac{1}{2}m_{water}v_{water}^2 . \tag{4.9}$$

Consider two cases: a smaller paddle blade with an area represented by A_1, and a bigger paddle blade with an area represented by A_2. We assume that with each stroke the paddle pushes an idealized plug of water with a volume covering the blade's area, extending a distance T behind the power face.[19] At the end of the stroke the paddler moves a mass of water equal to the volume of water (the blade area A times the thickness T) times the density of water ρ:

$$m_{water} = A \times T \times \rho . \tag{4.10}$$

For the "smaller blade" paddle, the total system kinetic energy at the end of the stroke is then

$$E_1 = \frac{1}{2}m_{hull}v_{hull}^2 + \frac{1}{2}\rho A_1 T v_{water,1}^2 . \tag{4.11}$$

For our "bigger blade" paddle, the total system kinetic energy at the end of the stroke is

$$E_2 = \frac{1}{2}m_{hull}v_{hull}^2 + \frac{1}{2}\rho A_2 T v_{water,2}^2 . \tag{4.12}$$

The ratio of these two energies is

---

19 This is merely a simplifying assumption that lets us compare the impact of blade size, as the detailed hydrodynamics of blades is a good deal more complex.

$$\frac{E_1}{E_2} = \frac{m_{hull}v_{hull}^2 + \rho A_1 T v_{water,1}^2}{m_{hull}v_{hull}^2 + \rho A_2 T v_{water,2}^2} . \tag{4.13}$$

For comparison, we'll assume that the hull's kinetic energy is the same for both blade sizes, i.e., that the hull is brought to the same speed using either paddle. This means that the first terms in both the numerator and denominator of equation (4.12) are the same. The difference lies only in the kinetic energies of water moved by the small and large blades, respectively. The ratio of these energies is

$$\frac{E_{water,1}}{E_{water,2}} = \frac{A_1 v_{water,1}^2}{A_2 v_{water,2}^2} . \tag{4.14}$$

Using the momentum equation (4.5) for each scenario allows us to solve for $v_{water,1}$ and $v_{water,2}$ in terms of the blade areas. After a little bit of algebra, the ratio of energies (13) simplifies to

$$\frac{E_{water,1}}{E_{water,2}} = \frac{A_2}{A_1} . \tag{4.15}$$

QED.

Further Reading

Halliday, Resnick, and Walker, *Fundamentals of Physics, Fifth Edition*, John Wiley & Sons, Inc. (1997).

CHAPTER 5

Roll Your Boat

Introduction

So, ever wonder why your hull flips? Yeah, me too. Usually after I'm home, dried off, and warming up under an afghan.

The obvious answer to "Why does a hull capsize?" is that you heeled it too far to one side. But how far is too far? What design factors influence roll stability? Why are some hull shapes more stable than others? And why are some hulls "twitchy"? Each of these questions has an answer found in physics.

In this chapter we'll consider the problem of roll stability in general terms. Why general? Because hulls are so different. An Old Town 'Tripper' canoe feels a *lot* different than an ICF kayak; they have radically different stability characteristics. A more detailed analysis requires hydrostatic models that are hull-specific. You'll also need naval architecture design and analysis software. Rather than do a deep dive for every hull that might be of interest we'll instead look at a few common hull shapes to provide insight into our questions.

Hulls and Restoring Forces

Our story begins more than 2,000 years ago, with an engineer named Archimedes.

There is an apocryphal tale of Archimedes running naked down the streets of Syracuse yelling "Eureka!" after discovering a method for determining whether a crown was made of gold. His method relied on water displacement to determine the crown's volume, a scale to determine its weight, and calculating its density from the ratio of weight to volume.

Archimedes then compared this result with the known density of gold. He supposedly figured out how to do this by observing the amount of water displaced in his bath after climbing in. Historians doubt that this ever happened; it doesn't make sense from an engineering standpoint in light of the required precision for bathwater level measurement. But seeking inspiration while chilling out and taking a bath does sound appealing.

Factually, in his treatise *On Floating Bodies*[20] this ancient Greek mathematician, scientist, philosopher, and engineer showed that an object floating in a liquid displaces a mass of fluid with a weight equal to that of the object. Once floating, the object continues to float; with all other factors constant it neither rises nor sinks after achieving equilibrium. As we now know from Isaac Newton this equilibrium occurs because the there is a *balance of forces*, equal in magnitude and in opposite directions so that their sum equals zero. This is an embodiment of Newton's 3rd Law. The floating object's weight "points" downward because of the acceleration due to gravity.[21] The buoyancy force must have equal magnitude and be directed upward to satisfy equilibrium. This is because force is a vector, having both size and direction. For the weight and buoyancy vectors to cancel they must align, point in opposite directions, and have equal magnitudes.

While force equilibrium may seem abstract, we experience it all the time. You are part of a balance of forces right at this moment – and you're probably not aware of it. Your weight can be conceptualized as a vector acting through your center of mass (more on that in a moment), because your weight – a force – has magnitude and direction. The direction is, of course, downward, toward the center of the Earth.[22] If your weight vector pointed up, you'd float away; if it pointed sideways, you'd accelerate right off the couch and into the cat, a wall, or whatever else stands in your way. Recall that Newton's 2nd Law requires that *unbalanced forces accelerate bodies*.

Your weight vector points downward, yet we note that you aren't accelerating through the couch into the basement. Your weight is therefore not an unbalanced force. Newton's 3rd Law tells us that there must be an equal and opposite force to satisfy static equilibrium. What is this force? Well, strange as it may sound, your couch (or chair, or whatever you're sitting or perhaps standing on as you read this) is providing this balancing force.[23] If you follow the entire chain of force transmission, then a similar equilibrium exists between your chair, couch, or feet and the floor. Then between your house and its foundation upon the Earth. This is the basis of statics, and much of civil engineering.

As to where these force vectors are located or how they are directed, an object with mass distributed over space can be idealized as a *point mass* of zero dimension coincident with the object's *center of mass*. The center of mass is the location where you can apply a force to accelerate an object that does not cause it to rotate. Bear in mind that a point mass is

---

20 https://archive.org/details/worksofarchimede00arch/page/262/mode/2up, pp. 263-300

21 Which is why you can leave your bathroom scale on the floor rather than bolt it to a wall or the ceiling; your weight "points" downward.

22 We'll assume that you're not floating weightless on the International Space Station.

23 Each is both cause for and effect from the other; give it some thought, and you'll soon come to the same conclusion. Those of you familiar with "dependent co-arising" are no doubt smiling.

merely a concept, a useful idealization. It's helpful for representing the static and dynamic properties of an object without accounting for what happens to each of its parts.[24]

The location of your body's center of mass will vary depending on how your various parts are arranged. For simplicity, we'll assume that you are symmetrical side-to-side. As shown in Fig. 5.1, my wooden stunt double has a center of mass in the sagittal plane located below its navel when in a standing posture. When seated the body's center of mass is located near its solar plexus; these are the locations where the figure balances on the tip of my finger. The stunt double doesn't rotate or fall off my fingertip because the weight vector through its center of mass and the reaction force exerted by my finger are aligned. The center of mass moves upward within the trunk when seated, and our legs extend forward.

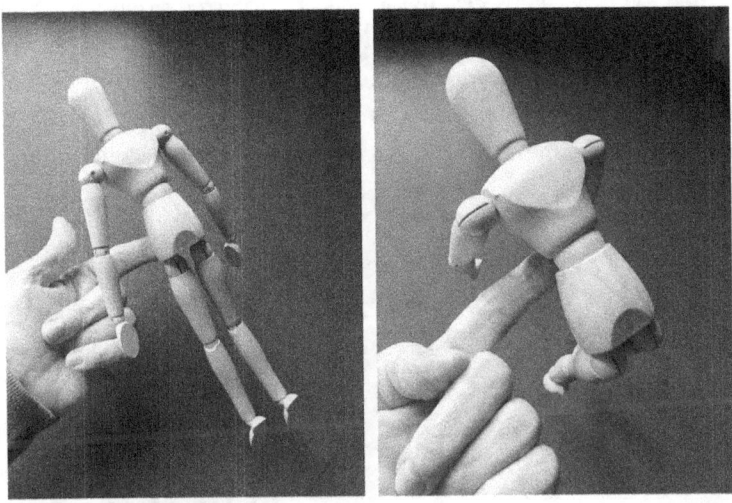

Figure 5.1: The body's center of mass for two postures.

Archimedes posited that the buoyancy force acted upward for floating objects – it is after all a buoyant force, not a sinking force – can be idealized as acting through the center of mass of the displaced "hole" created in the water by the object. Let's consider a specific example: a cylindrical log floating in the water. As shown in Fig. 5.2's cross-section, the log displaces a particular volume of water. The weight of the displaced water equals the weight of the log. The buoyant force is directed upward through the centroid of the displaced water mass. This centroid is called the *center of buoyancy* and is labeled **B** in the figure. Fig. 5.2 does not suggest that water has somehow seeped into the log. Fig. 5.2 only represents the equivalence of these two scenarios if the "virtual water" in the right-hand side drawing has negative weight, i.e., an upward pointing force vector. It makes the analysis easier.

---

24 For readers with a philosophical bent this reflects science and engineering's ability to tell us about what things *do* while being silent about what they *are*. Science and engineering are functional, not ontological. As Newton wrote concerning his model of gravity, "I have not as yet been able to discover the reason for these properties of gravity from phenomena, and I frame no hypotheses."

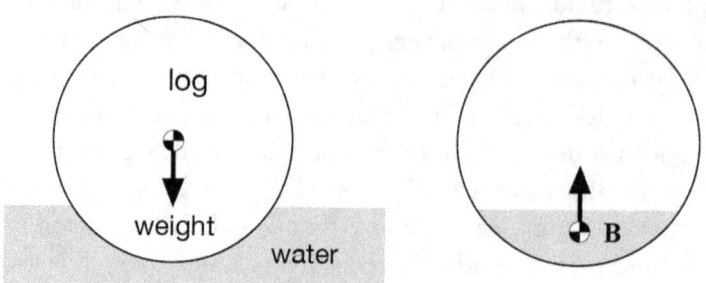

Figure 5.2: Log in water and buoyant force location.

Since our idealized log is perfectly circular, it has no preferred orientation. In other words, you can rotate the log and it will continue to float; its center of gravity does not move. The log's weight vector will always be opposed (and canceled) by the buoyancy force vector since they never misalign.

Let's make things more interesting by putting an idealized paddler in an idealized hull with a semi-circular cross section. This scenario is depicted in Fig. 5.3.

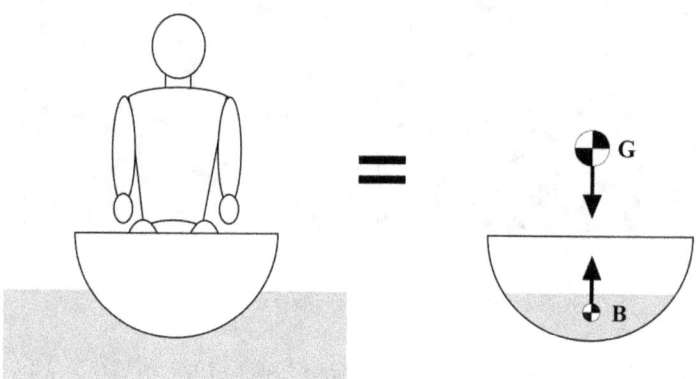

Figure 5.3: Paddler in hull with a semi-circular cross section.

The center of mass **G** (also called the *center of gravity*) is where the combination of the hull, paddler(s), and gear can be represented as a single point mass. Their combined weight is represented as a single downward-pointing weight vector acting through **G**. As suggested in the right-hand side of Fig. 5.3 the weight vector and the buoyancy force vector are aligned and cancel each other out. The lengths of the two force vectors, corresponding to force magnitude, are equal. Since there are no unbalanced forces the hull and paddler will remain upright and floating.

But for anyone who's paddled a hull like this, it gets a lot more interesting when you heel the boat – as if climbing into that hull and trying to keep it vertical wasn't exciting enough. Heeling is represented in Fig. 5.4, using only point masses for clarity. When the circular hull heels, the displaced volume of water holds the same shape and position: a circular segment subtended by the hull's surface and a chord at the waterline. As a result, the center of buoyancy **B** for the semi-circular hull never moves side-to-side until the gunnels submerge. By contrast the center of mass of the hull plus paddler moves toward the heeled side. This results in unbalanced forces. And as we know from Newton, unbalanced forces cause masses to accelerate. Which for our paddler means the hull will roll, the roll will accelerate, and unless they brace, it's trout scouting time.

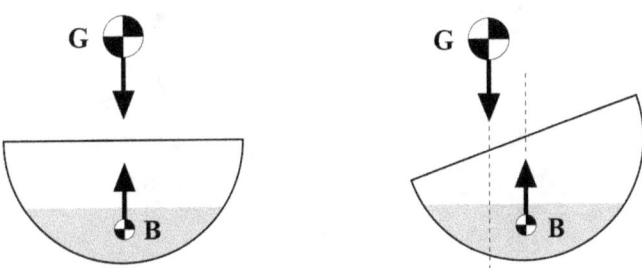

Figure 5.4: Semi-circular hull, vertical (left) and heeled (right).

For comparison, consider the V-bottomed, hard-chined hull depicted in Fig. 5.5. It looks a little bit like a 3x27 USCA Pro Boat. Again, the hull and paddler are idealized as a point mass located at the center of mass **G**, along with a counterbalancing buoyant force at **B**. When the hull is vertical the respective force vectors cancel each other out, the hull floats, and everyone's happy.

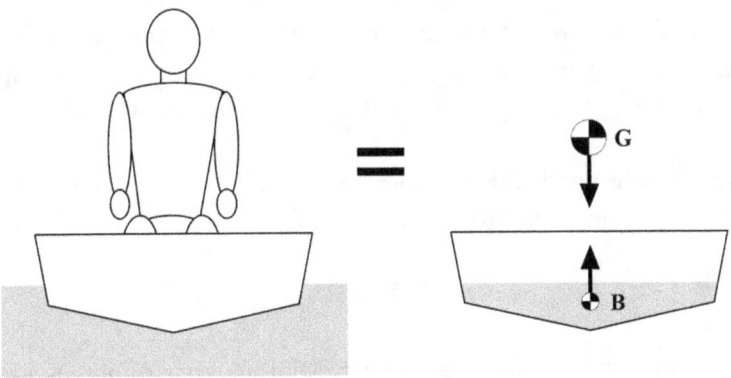

Figure 5.5: "Pro Boat" and point mass equivalents.

When this hull heels, the center of mass **G** moves toward the heeled side as before. But because of the hull's cross-sectional shape something interesting happens with the center of buoyancy: it moves toward the heeled side, as depicted in Fig. 5.6.

Again, the forces are unbalanced and there is a net moment acting on the hull. But this time the buoyant force is *outside* of the weight vector. The moment created by the unbalanced forces will rotate the hull *away* from the heeled side rather than toward it. In this case the buoyant force acts as a *restoring force*. Then, as the hull rights itself the buoyant force moves back toward the keel line, always staying outside (or collinear with) the weight vector until the hull is vertical. Because the hull is symmetric side-to-side if the hull overshoots vertical, the buoyant force will move toward the other side, staying outside the weight vector and thus righting the hull

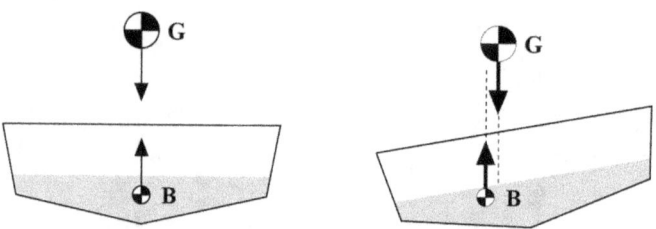

Figure 5.6: "Pro Boat" vertical and heeled.

Now we've all flipped a boat or two. You can feel the restoring force help the hull set up at a certain lean angle and stay there. Beyond this point, especially for racing hulls or highly elliptical hulls, the restoring force moves inside of your weight vector, and you either brace or swim. The reason hulls do this is because after a certain heel angle, you either put the gunnels below the waterline and swamp, or the center of buoyancy can no longer move outside of the center of mass at higher angles of heel. This effect can be characteized using a quantity called the *metacenter*.

The metacenter **M**, depicted in Fig. 5.7, is the intersection point of the buoyancy force vectors and the hull's centerline for various roll angles. The center of buoyancy follows an arc as the hull leans. For small roll angles the metacenter does not move up or down the centerline; for these angles the arc subtended by the various centers of buoyancy describes a circle.

If we assume that the paddler does not lean as the hull rolls, the center of gravity remains on the centerline. For small hull roll angles ϕ,[25] the *righting arm* is defined as

$$\overline{GM}\sin(\phi).\tag{5.1}$$

The righting arm geometry is illustrated in Fig. 5.8; the overbar notation in equation (5.1) signifies the distance between **M** and **G**. The righting arm is the moment arm upon which the buoyancy force "pushes" to right the hull. The greater the metacenter height is compared to the center of mass, the greater the righting arm, resulting in a larger restoring torque for righting the hull.

---

25 The limits of which depend upon the hull's underwater shape.

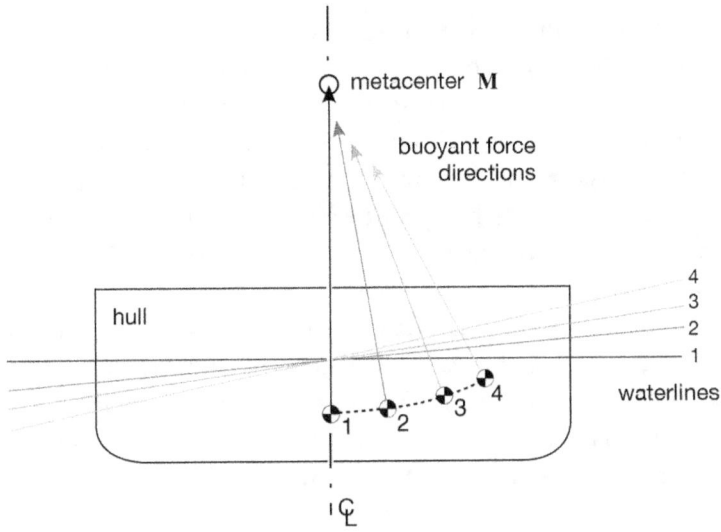

Figure 5.7: Metacenter location and arc of buoyancy force vectors.

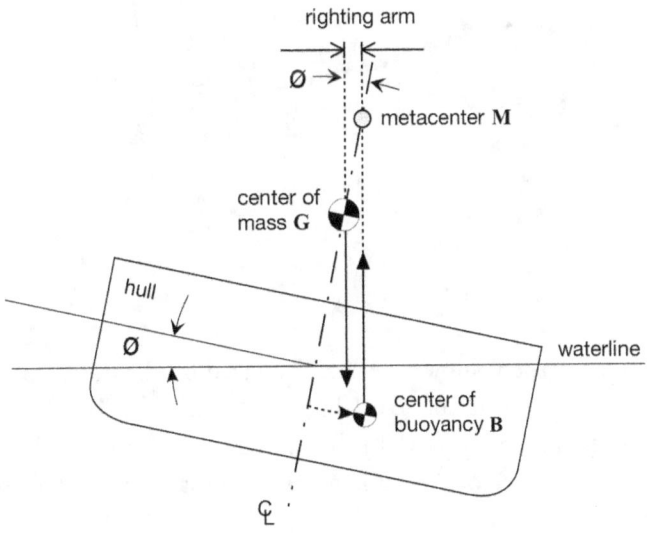

Figure 5.8: Righting arm definition.

As roll increases the metacenter moves downward because the arc described by the moving buoyancy centers curves upward. This is because the buoyancy vector is everywhere perpendicular to this arc; its intersection with the centerline will move downward. When the metacenter is at or very near the center of gravity then the hull may have a permanent list due to the lack of an adequate righting arm. This list can occur when a hull fills with water. And when the metacenter moves below the center of gravity the hull will capsize absent any additional restoring force, such as a paddler throwing a brace.

This restriction on the allowed center of buoyancy movement depends upon the shape of the hull. Because the hull sides curve upward, the shape of the subtended mass of water stops changing all that much after a certain heel angle. And since the buoyant force acts through the center of mass of this "hole" in the water while its shape doesn't change (much), the side-to-side location of **B** now changes little with roll angle. And voila! Trout scouting. The center of mass **G** has moved outside of the center of buoyancy **B**.

You can counter a hull's tendency to capsize by extending its cross-section width to infinity. However, infinitely wide boats don't exist. You can install sponsons, which you'll find (for example) on Sportspal canoes; this will help a bit if you're leaning over the gunnels to pull in a fish. You can learn how to throw a brace – a temporary fix that assumes you're on the correct side to counteract the roll.[26] Or, you can install an outrigger, which acts as a dependable restoring force – at least to the side with the ama – as illustrated in Fig. 5.9.

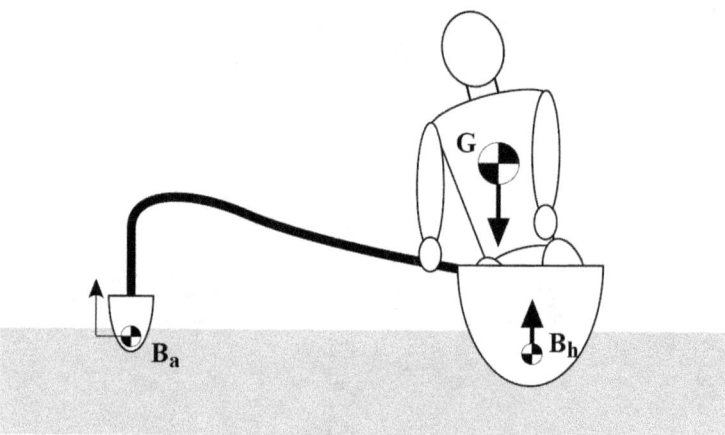

Figure 5.9: Outrigger hull force balance.

As illustrated in Fig. 5.9, the volume of water displaced by the outrigger's ama provides a second buoyancy force, B_a, in addition to the hull's buoyancy force B_h. This second buoyancy force acts over a longer moment arm than the moment arm between the center of gravity and the metacenter and will right the hull. With the ama in the water, the outrigger will be stable in roll to its side.

One last thing about the interaction between **G** and **B**: it's better to be a shorter paddler. Fig. 5.10 shows the centers of mass for two paddlers of different heights; the difference is exaggerated to illustrate the effect more clearly. For a given angle of heel the taller paddler's center of mass, and thus weight vector, are closer to the center of buoyancy than the short paddler's center of mass. The taller paddler is closer to the stability limit. There is more margin for the shorter paddler because the metacentric height starts higher to their center

---

26 A successful brace also depends on paddle blade shape, and employing the correct blade face. Think low brace.

of gravity compared to a taller paddler. The hull can heel further before the shorter paddler exceeds the stability limit.

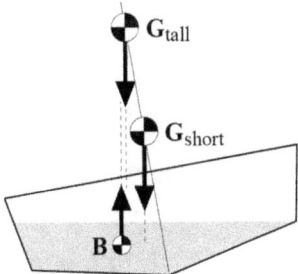

Figure 5.10: Movement of G for two paddler heights.

The cross-sectional shape of the canoe below the waterline also influences how the canoe "feels" about the vertical. A shallow arch hull will have a center of buoyancy that moves smoothly with heel, and the metacentric height is located well above **G**. A V-hull will have a center of buoyancy that moves abruptly with heel. The metacentric height drops rapidly down the hull center line from the moment the hull is first heeled. Both hulls may eventually set up after a certain amount of lean. But because **B** initially moves abruptly in a V-hull, and the metacentric height moves abruptly toward **G**, it will tend to feel "twitchy" about the vertical. Because... physics.

For those of us who like running whitewater or paddling in wind and waves, we can now tell our friends why we flipped. "It was physics, not me!"

Take-Aways

- The water displaced by a hull and its contents gives rises to a buoyancy force. This force is directed upwards, and precisely equals the weight of the hull and its contents. This principle, which embodies Newton's 3rd Law, was first discovered by Archimedes.

- Buoyancy acts as a restoring force when most hulls roll. This is because the center of buoyancy moves toward the direction of roll due to the hull's shape, creating a righting arm that keeps the hull from rolling further. Unfortunately, hulls with a circular cross-section don't perform that way; their center of buoyancy never lies outside their center of mass.

- Absent other external forces, hulls capsize when the center of buoyancy moves inside the center of mass of the hull and its contents. This corresponds to the metacenter moving below the center of gravity. It's that simple.

- If the metacenter is above the center of gravity, the righting arm between the center of buoyancy and the center of gravity has a stabilizing effect to bring the ship back to its normal position.

- If the righting arm is displaced below the center of gravity, the hull will lose its roll stability and capsize.

CHAPTER 6

About the Paddle

Introduction

In Chapters 1 through 5 we focused on the hull: Why it floats, roll stability, and impediments to motion in the form of friction, form, and wave drag. We now turn to the paddle, without which there would be no paddling.

Paddles embody a wide variety of shapes and sizes, as illustrated in Fig. 6.1. Native paddlers in Greenland and Northern Labrador developed narrow double-bladed paddles to propel their *qajaqs*. Micmac paddlers carved elegant beavertail-shaped canoe paddles with blade width bounded by the diameter of local trees. Passamaquoddy paddle makers employed ottertail blade shapes, which are both elegant and very quiet in use.

The common elements of all the paddles shown in Fig. 6.1, and subsequent paddle designs, are a shaft and one or two blades. The shaft links the paddler to the propulsive element, i.e., the blade. And how does the blade create propulsion? Recall our thought experiment from Chapter 2, where we considered the form drag created by your hand as you wave it side-to-side. When your palm is flat and aligned with the direction of motion, the air exerts less force on your hand than when the palm is perpendicular to the flow. Now try the same thing with a paddle. If you slice it through the water with the blade aligned to the direction of travel, you won't be going far. If instead, you orient the blade perpendicular to the direction of travel, you'll notice a lot of resistance to its movement. By using the blade in this way, the paddle creates a drag force. For our hulls, drag is an ever-present nemesis.

Figure 6.1: A passel of paddles: 1. Inland Chukchi, 2. Kodiak, 3. Norton Sound, 4. Bering Sea Yuit, 5. Norton Sound, 6. MacKenzie River Delta, 7. Copper Inuit, 8. Caribou Inuit, 9. Labrador, 10. Polar Greenland, 11. Upernavik Greenland, 12. Illorsuit Greenland, 13. 17th C. West Greenland, 14. 17th C. West Greenland, 15. West Greenland (1700s - 1800s), 16. West Greenland (1700s - 1800s), 17. West Greenland 17th C, 18. East Greenland.

For paddles, the force they generate through drag is their *raison d'etre*. One might even think of a paddle as a hydrodynamically inefficient hull and a hull as an inefficient paddle. So then, why does any paddle moving through the water propel a canoe, kayak, or SUP in the first place? How can we make our paddle strokes more efficient? Are "bent shaft" paddles more efficient at moving a canoe, and if so, why? Is there an optimum bend angle, and if so, what does it mean for the angle to be optimal?

To answer these questions, we'll employ a simple physics model of how a paddle generates force to quantify efficiency and optimality. We'll use a data-driven biomechanical model of shaft movement to compute paddle force as a function of blade angle, i.e., force over time. This model shows that we maximize peak paddling force and impulse – the time integral of force – if the blade is perpendicular to the water at the instant that the relative velocity between the paddle and the hull is maximal.

We'll begin with a little physics.

Ein Bisschen Physik[27]

While Isaac Newton wasn't a paddler, perhaps he was inspired by punters on the River Cam when he developed his laws of motion. The 2nd Law states that forces acting on a body will accelerate it in inverse proportion to its mass.[28] For a hull, paddler, and paddle with a combined mass M, the 2nd Law of Motion is

$$M \frac{d}{dt} v_{hull} = F_p - F_D , \tag{6.1}$$

where d/dt denotes the time rate of change of the hull's velocity[29] v_{hull}, F_p represents the propulsive force exerted via the paddle, and F_D is the total drag force acting on the hull in opposition to the propulsive force. The right-hand side of the equation is the *net* force on the hull; the sign difference between the propulsive and drag forces reflects how these oppose each other. Except for the mass M, all quantities in equation (6.1) are functions of time.

We've discussed the contributors to the drag force F_D in Chapters 1 through 3. Ultimately the components of drag – form, friction, and wave – are properties of nature, motion, and the hull design. Given a particular hull, we're left to consider how the propulsive force F_p produces hull speed and how to optimize paddle force.

The paddle's propulsive force is expressed in terms of the *relative velocity* between the hull and the paddle,

$$F_p = \frac{1}{2} \rho C_d A \left(v_{hull} - v_{paddle} \right)^2 , \tag{6.2}$$

where ρ ("rho") is the density of water, A is the area of the paddle blade, and v_{paddle} is the velocity of the paddle blade's center in an inertial reference frame relative to the Earth.[30] The choice of reference frame is critical since it is the relative velocity between hull and paddle that contributes to the propulsive force. The hull could be moving swiftly downstream, and the paddle along with it if you're not paddling; in that case, you're exerting no propulsive force, yet the velocity of the paddle by itself could be considerable. Finally, since velocity is a vector (directional) quantity, the velocities here are expressed in the direction of travel.

The term C_d is the paddle's drag coefficient, which is the dimensionless quantity that characterizes the blade's resistance to motion through the water. The drag coefficient is a property of the blade design and blade angle. In equation (6.2), we see that the non-dimensional drag coefficient is distinct from the blade area A. The propulsive force F_p is also a function of the shaft angle θ ("theta"),[31] defined in Fig. 6.2. The total pressure distribution

---

27 Well, if Mozart got 'Eine Kleine Nachtmusik," why can't physics?

28 The applied force changes the object's momentum. But if the mass is constant, then the equation simplifies to this familiar form. See Chapter 0's Extra Credit section for details.

29 This time rate of change of velocity is acceleration, and d/dt is a derivative. Just dropping a little calculus.

30 You can reference the velocities to the water if the current is steady and non-zero. This introduces a coordinate shift called a Galilean Transformation since the two reference frames differ only by constant relative speed.

31 Since the blade is rigidly attached to the shaft, you know the blade angle if you know the shaft angle.

over the blade face yields a net force perpendicular to its power face – as you'll recall, force is a vector. Now paddlers know that the angle of the power face changes over a stroke.

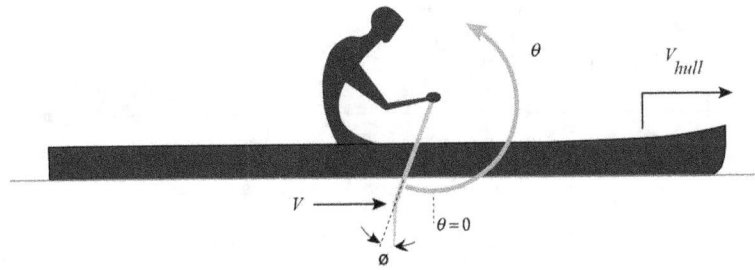

Figure 6.2: Coordinate definitions.

Since we're interested in the propulsive force in the direction of travel, and will assume that the center of pressure on the blade's power face during a stroke does not move up or down, the angular dependence of the paddle's drag coefficient can be written as

$$C_d = C_{d0} \cos\theta. \tag{6.3}$$

C_{d0} is the drag coefficient measured when the blade's power face is perpendicular to the water's surface (e.g., $\theta = 0$), and $\cos(\bullet)$ is the cosine function. The cosine ensures that we compute the force *in the direction of travel* at any instant/angle, rather than force perpendicular to the blade's power face. To reinforce this, we'll hereafter refer to it as the *direction cosine* and the component of the paddle force in the direction of travel the *propulsive force*. So, if we know the shaft angle over time and the relative velocity over time, we can compute the propulsive force function for a given paddle area A and drag coefficient C_{d0}.

Modeling Relative Velocity and Shaft Angle Over Time

A sequence of events repeats during every forward paddle stroke, as illustrated in Fig. 6.3. These are (1) the catch, where the paddle blade enters the water, positioning it for the next part of the sequence; (2) the power phase, where the paddler pulls on the shaft to move the hull forward; (3) the exit, where the blade leaves the water; and (4) the recovery, where the paddle bade is sliced/swung forward in preparation for the next stroke. We'll assume that the entry and exit phases do not contribute to the hull's forward motion, and the recovery does not contribute to propulsion. Consequently, we'll focus on the power phase and how the shaft angle and relative velocity vary over time there.

We'll use a mathematical model for the shaft angle and relative velocity derived from experimental data obtained by Nicholas Caplan. Caplan employed high-speed video cameras to measure the shaft angle over time, as well as the relative hull/paddle velocity, of an elite outrigger paddler using a straight shaft paddle.[32] By using his data, we are assuming that the

---

32 Nicholas Caplan, "The Influence of Paddle Orientation on Boat Velocity in Canoeing," *International Journal of Sports Science and Engineering*, **3**(3), pp. 131-139 (2009).

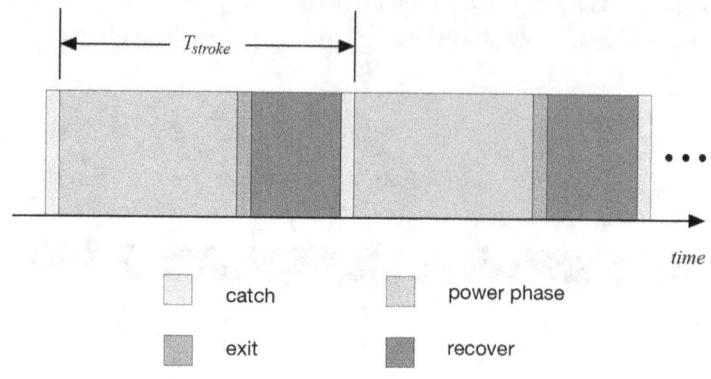

Figure 6.3: Phases of the paddle cycle over time.

stroke mechanics employed by this outrigger paddler applies to other paddling disciplines, using either straight shaft or bent shaft paddles. That's a big assumption. However, the force curves computed using our model have the same characteristics as those of sprint kayakers.[33]

Our model applies from paddle entry to exit since we generate no blade propulsive force during the other phases. For a double-bladed paddle, the model applies to each blade individually. The second blade merely repeats the same motions with the catch delayed by the stroke cadence.

We can approximate Caplan's experimental data rather well using simple mathematical functions. The measured relative velocity is be represented by

$$v_{hull} - v_{paddle} = -3.5 \cos\left(\frac{\pi t}{0.4} - \frac{\pi}{2}\right) - \frac{t}{0.4} \quad \text{meters/sec} \cdot \tag{6.4}$$

This approximation, plotted in Fig. 6.4, is valid from the time of the stroke's catch at $t = 0$ seconds until the exit at $t = 0.4$ seconds. The relative velocity is negative because the magnitude of the paddle velocity is greater than the magnitude of the hull velocity during the power phase. The negative relative velocity also indicates that the paddle velocity is directed toward the stern, consistent with momentum conservation. The relative velocity varies over time, with a magnitude that reaches its maximum value a bit after mid-stroke, as shown in the figure.

We approximate the shaft angle as a linear function over time. The angle ranges from a minimum of -30 degrees at the catch to a maximum of +50 degrees at the exit. The shaft angle θ is then

$$\theta(t) = 200t - 30 \quad \text{degrees for } 0 \leq t \leq 0.4 \text{ sec} . \tag{6.5}$$

33 Beatriz B. Gomes, Nuno V. Ramos, Filipe A.V. Conceição, Ross H. Sanders, Mário A.P. Vaz, and João Paulo Vilas-Boas, "Paddling Force Profiles at Different Stroke Rates in Elite Sprint Kayaking," *Journal of Applied Biomechanics*, **31**, pp. 258-263 (2015).

This angular dependence is plotted in Fig. 6.5. Since the curve is linear, the shaft angle versus time doesn't have an inflection point, just minimum and maximum angles at the catch and exit.

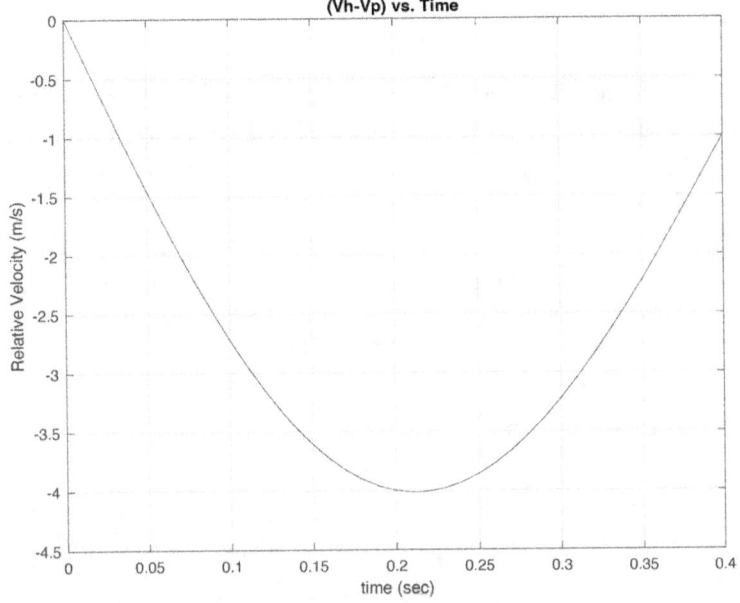

Figure 6.4: Relative velocity from catch to exit.

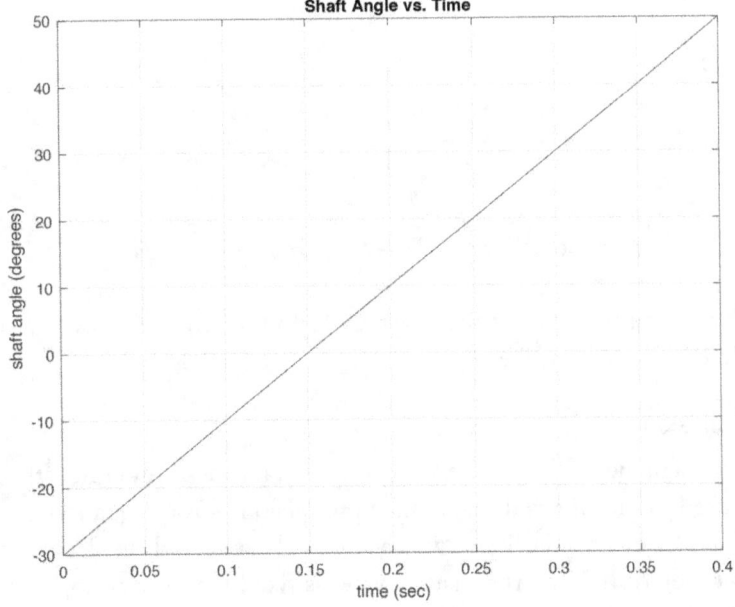

Figure 6.5: Shaft angle from catch to exit.

Given this model, we can now compute paddle force over time using the following parameters:

- Blade area: 8"x12" = 96 *sq. in.* = 0.0632 *m²*
- Water density: 998.57 *kg/m³*
- C_{d0}: 1.28 (flat plate)

The resulting paddle force over time appears in Fig. 6.6. Note how there is a short "shelf" in the force curve at the start of the power phase. This shelf reflects the cosine dependence on shaft angle and the squared dependence on the relative velocity. The force profile has a slightly skewed cosine shape and ends with a non-zero value. The skew reflects this paddler's stroke mechanics, where the blade exits while it's still loaded. This paddler, filmed at a national-level time trial, exerts a large blade peak force. For slower paddling cadences, we

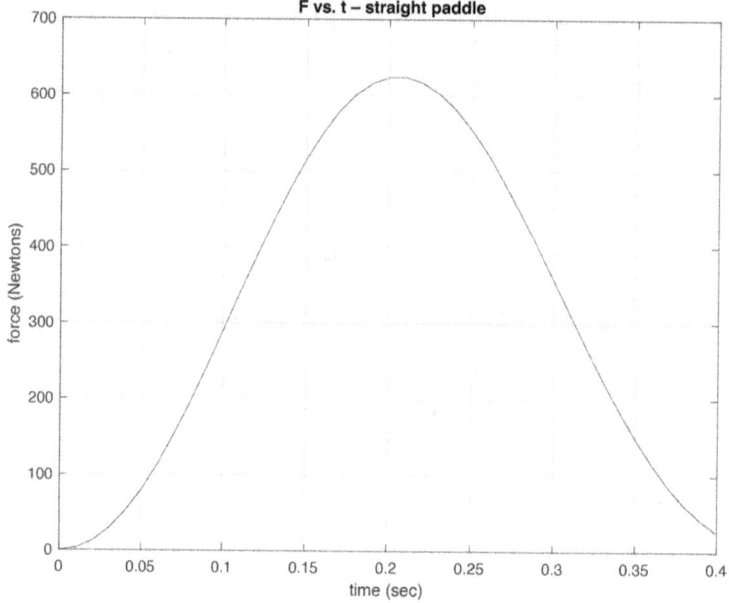

Figure 6.6: Force vs. Time for a straight-shaft paddle.

can stretch the time dependence of the shaft angle and relative velocities. And peak force will scale for less aggressive strokes.

Optimizing Propulsion

Now that we have a model of the paddle force, it's natural to ask how either stroke mechanics or the paddle itself might optimally influence propulsion. From equation 6.2, we see that paddle force depends on the density of water, the blade area, and the drag coefficient. The drag coefficient, determined by the blade shape, is fixed before you hop in the boat. So the optimization "knobs" we have available to us are the time-dependent shaft angle and relative velocity. We can see how these change over time in Fig. 6.7.

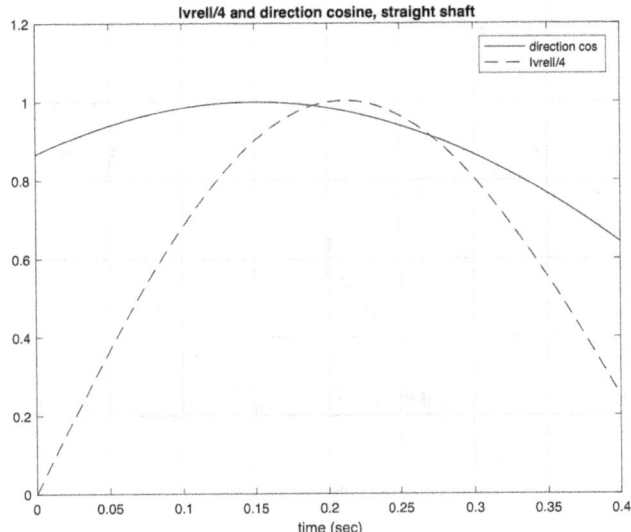

Figure 6.7: Relative velocity magnitude and direction cosine, straight shaft paddle.

Fig. 6.7 plots the relative velocity's magnitude versus time and the variation of the direction cosine for a straight shaft paddle. The magnitude of the relative velocity is divided by four (4) in this figure; this is purely for convenience so that the two curves fit in the same vertical scale. Since the shaft angle varies linearly over time, the figure shows that the straight shaft paddle reaches its maximum direction cosine value when the shaft (and hence the blade) are perpendicular to the water. This maximum occurs before the relative velocity reaches its maximum value, as seen in the figure. Engineers would conclude that the direction cosine "leads" the relative velocity for this paddle.

We know that the direction cosine and the relative velocity change over time, so to maximize their product and therefore maximize propulsion, we want them to reach their maxima simultaneously. Engineers call this kind of time relationship "in phase." From high school trigonometry, we know that the cosine function varies in magnitude from 0 to 1 and reaches a maximum when its argument is either zero or 180 degrees. Here, zero degrees corresponds to the shaft in the downward vertical position, and 180 degrees corresponds to upward vertical, like when doing a "paddle salute." So, we need to introduce a phase lag or delay in the paddle. We can do this by breaking it.

So-called bent shaft canoe paddles were developed in 1971 by famed Minnesota paddler and canoe racer Gene Jensen. He noticed that by bending a canoe paddle at the neck, where the shaft meets the blade, a paddler could move a hull faster than with an equivalent straight-shaft paddle. Jensen noted that paddlers using straight shafts "had to dig these enormous holes, and their paddles would really cavitate. They didn't seem to go as fast as they should." The rest, as they say, is history. You can scarcely go to a downriver or flatwater race and not see carbon fiber bent shaft paddles everywhere. And because of their efficiency, they've become rightfully popular with recreational paddlers as well.

We can capture this effect by generalizing equation 6.2 to include bend angle,

$$F_p = \frac{1}{2}\rho C_{d0} A \cos(\theta - \phi)\left(v_{hull} - v_{paddle}\right)^2. \tag{6.6}$$

In this equation, we've introduced an additional parameter ϕ ("phi") representing an offset to the shaft angle. Phi represents our "broken" paddle's *bend angle* between the paddle shaft and blade power face, as illustrated in Fig. 6.8.

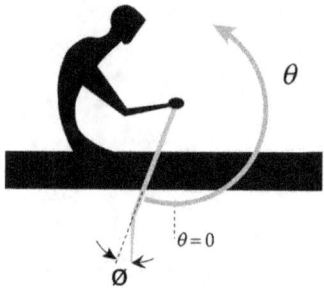

Figure 6.8: Bent shaft paddle geometry.

We varied the bend angle from 0 degrees (e.g., a straight shaft paddle) to 20 degrees in one-degree increments to determine which bend allowed the blade and relative velocity to be in phase, and thus maximize their product. In Fig. 6.9, you'll find a plot of the relative velocity's magnitude versus time, as well as the variation of the direction cosine for two blade angles. The magnitude of the relative velocity has been divided by four (4) in this figure. The direction cosine is plotted for a bend angle ϕ of zero (0) degrees (no bend) and twelve (12) degrees. Since the shaft angle varies linearly over time, the figure shows that the 12-degree bent shaft paddle maximizes its direction cosine when the relative velocity hits its maximum. The bent shaft's blade angle is in phase with the relative velocity; the blade is perpendicular to the water's surface when the relative velocity is maximum. As a result, we've maximized propulsion by breaking the paddle.

You'll find a plot of the propulsive forces corresponding to a straight shaft and a bent shaft paddle in Fig. 6.10, computed using the equation above. The paddle force for both cases starts at zero since, at the catch, you haven't started to pull. The bent shaft's force profile – the variation in force over time – lags the straight paddle; the bend introduces a phase lag in the force,[34] as expected. The bent shaft paddle hits a higher peak force than the straight shaft paddle due to the temporal phase alignment of the angle of the blade face and the relative velocity. Further, it reaches this peak a bit later than for the straight shaft paddle. The relative velocity reaches maximum magnitude a bit later than the power phase's midpoint, and the bent shaft's blade is perpendicular to the water at that moment.

So, to maximize propulsion, your paddle blade should be perpendicular to the water when the relative velocity between the blade and the hull is greatest. Why not just cut off the stroke at that point? Wouldn't that make the hull move fastest?

---

34 What would Obi Wan say?

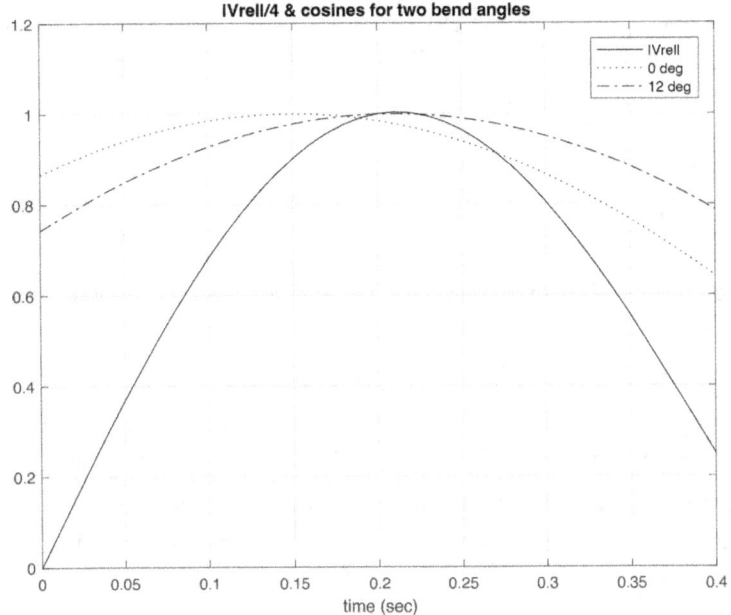

Figure 6.9: Relative velocity magnitude and direction cosine from catch to exit.

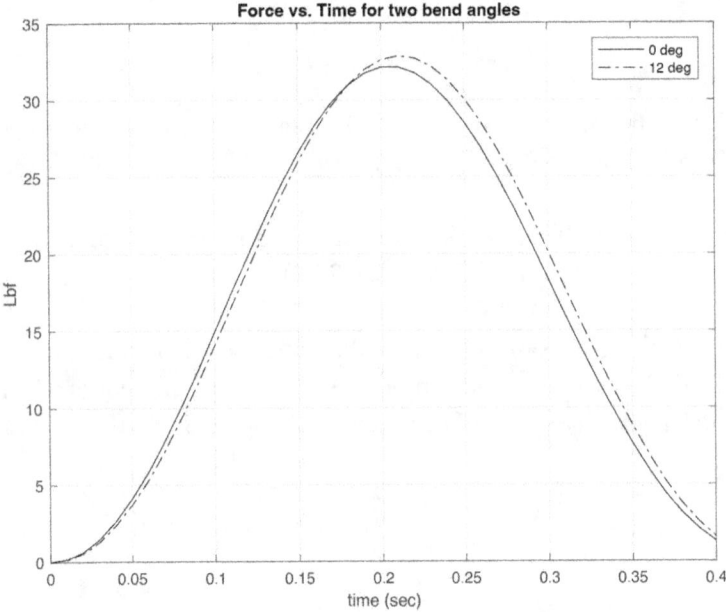

Figure 6.10: Force vs. Time for straight- and bent-shaft paddles.

Short answer: no.[35] But great question! To arrive at this conclusion, you need to sum all the paddle's propulsive force contribution over the entire stroke, not just the peak force.

---

35 With a caveat: Ultimately, you must consider stroke cadence and the relative amount of time you spend in the power phase vs. the recovery phase. We'll address this in Chapter 7.

This quantity is what physicists and engineers call *impulse*. Impulse produces a change in linear momentum, which we learned in Chapter 4 determines cruising speed. We'll consider impulse in detail in the next chapter.

Take-Aways

- Paddles are essentially miniature barn doors that we sweep through the water to propel our craft.

- Paddles create a paddle force primarily because of form drag, and secondarily from friction drag.

- The paddling stroke is divided into four phases: catch, power, exit, and recovery.

- A paddle's propulsive force is proportional to its blade area, as well as the squared relative velocity between the blade and hull during the stroke's power phase.

- You can optimize propulsion by tuning your technique so that the blade's power face is perpendicular to the water when the relative velocity between the blade and hull is maximum.

- Bent shaft canoe paddles address the "phase lead" of their straight shaft cousins, where, depending on the bend angle, the blade can be perpendicular to the water's surface when the relative velocity is maximum.

Further Reading

Nicholas Caplan, "The Influence of Paddle Orientation on Boat Velocity in Canoeing," *International Journal of Sports Science and Engineering*, **3**(3), pp. 131-139 (2009).

D. Morgoch and S. Tullis, "Force analysis of a sprint canoe blade," *Proc. IMechE*, **225**, Part P: *Journal of Sports Engineering and Technology* (2011).

Beatriz B. Gomes, Nuno V. Ramos, Filipe A.V. Conceição, Ross H. Sanders, Mário A.P. Vaz, and João Paulo Vilas-Boas, "Paddling Force Profiles at Different Stroke Rates in Elite Sprint Kayaking," *Journal of Applied Biomechanics*, **31**, pp. 258-263 (2015).

CHAPTER 7

Impulse

Introduction

I have periodically had discussions – well, perhaps arguments – with paddlers and runners about ways to parameterize our sports. Aside from race finishing time or cruising speed, what are the "best" things to measure so we can understand the impact of a particular aspect of paddling or running? And are these metrics actionable?

Running is my favorite case in point. If each stride is longer and the number of strides per minute (aka, cadence) is higher, we intuit that we run faster, which is true. Stride length and cadence are easy to wrap our heads around. These quantities can be measured using wrist-mounted sports computers and foot pod speed/distance sensors. But longer strides *result from* better underlying running mechanics; they are not the cause. Cadence and stride length are a step removed from the underlying causes.

Consider Fig. 7.1, from U.S. Patent Number 6,018,705 awarded to Gaudet *et al.* on January 5th, 2000. This figure plots pace vs. foot contact time – the time your foot is on the ground – for running and walking. The plot comprises two straight trend lines, with an abrupt transition between the two. The left line and its slope relate foot contact time with running pace, while the right line and its slope relate foot contact time with walking pace. The transition reflects changing stride mechanics as we move from walking to running.[36]

---

36 When we walk, one foot or the other is on the ground. When we run, there are times when both feet are off the ground.

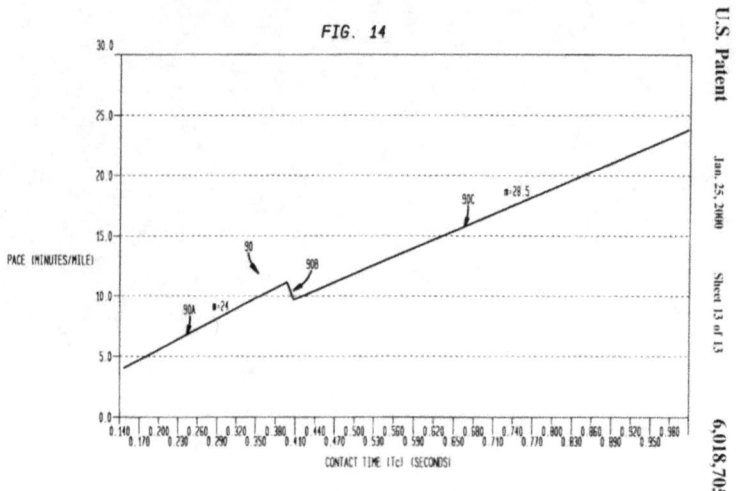

Figure 7.1: Pace vs. foot contact time for runners and walkers.

This plot suggests that neither stride length nor cadence plays a role in pace. And since pace is the inverse of running speed, we infer that neither stride length nor cadence plays a causal role in speed. Instead, a shorter foot contact time corresponds to a faster pace/speed. So, if you want to identify a trainable parameter to increase running speed, choose foot contact time.

What gives rise to shorter foot contact times? Further analysis of runners shows that speed is related to one's ability to exert large foot-ground forces quickly. As Clark and Weyland[37] note,

"Contrary to intuition, fast and slow runners take essentially the same amount of time to reposition their limbs when sprinting at their different respective top speeds. Hence, *the time taken to reposition the limbs in the air is not a differentiating factor for human speed.* Rather, *the predominant mechanism by which faster runners attain swifter speeds is by applying greater forces in relation to body mass during shorter periods of foot-ground force application.*" [emphasis added]

In other words, stride length is not causal or what some might think of as an "input" that makes us move. It is a result – an output.

When reading Clark and Weyland's paper, I recalled the physics principle relating impulse and momentum. When a bat hits a baseball, you change the ball's momentum by applying a force over a finite period of time. The same goes for driving a golf ball off a tee. The force-time input which changes the struck object's momentum is called *impulse*. Greater impulse entails greater distance traveled and greater speed.

Each paddle stroke is akin to hitting a golf ball as far as physics is concerned. You

---

37 Kenneth P. Clark and Peter G. Weyand, "Are running speeds maximized with simple-spring stance mechanics?", *Journal of Applied Physiology*, **117**, pp. 604 – 615 (2014).

change the hull's momentum by exerting a propulsive force over a finite time. But what about exerting "large" forces "quickly"? Is this stretching the running analogy too far?

In this chapter, we'll use impulse and momentum to characterize our propulsive inputs. We begin by carefully reviewing the sequence of events we'll call the *paddle cycle*. The paddle cycle helps us understand how propulsive force and time parameterize our paddling. Along the way, we'll identify actionable focuses for various stroke elements. The results apply to both competitive and recreational paddlers. The Extra Credit section shows that distance per stroke – the paddling equivalent of running's stride length – is a result rather than a cause.

Momentum and Impulse

As you'll recall from Chapter 4, we can describe the motion of a paddled craft in terms of its momentum p. Momentum is the product of the hull's mass, plus the mass of its occupants and gear, times its velocity,

$$p = mv . \tag{7.1}$$

In general, momentum is a vector (directional) quantity because velocity is a vector. The mass has no direction and is thus a scalar. In this chapter, we'll only consider velocity's magnitude in the direction of motion. This magnitude is the hull's speed. So, hereafter, 'v' signifies the hull's speed in the direction of motion. It is a scalar quantity.

Since the mass of the hull and its contents m is constant, a change in momentum Δp then corresponds to a change in its speed Δv,

$$\Delta p = m\Delta v . \tag{7.2}$$

This change in speed is illustrated in Fig. 7.2. With each stroke, in the water's reference frame, the hull speeds up during a stroke's power phase, then slows down because of drag forces until the next stroke. This cycle repeats itself stroke after stroke.

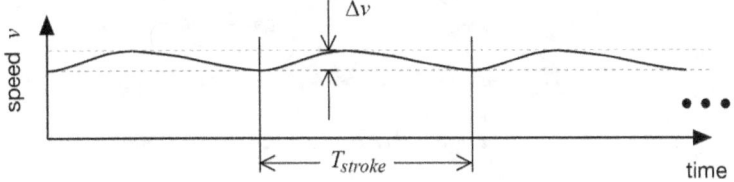

Figure 7.2: Hull speed variations over time (not to scale).

We can cast this change in speed in terms of momentum conservation. The paddle generates momentum opposite the direction of motion. Since the system's momentum is conserved, the hull speeds up. We can also cast the change in speed in terms of the force exerted by the paddle using Newton's 3rd Law of Motion. We'll draw from both in the analysis that follows.

First, recall that the hull's acceleration a is the time rate of change of its speed. Rather than writing this using the calculus operator called a derivative, we'll express the time rate of change in terms of the change in velocity Δv over a corresponding interval of time Δt,

$$a = \frac{\Delta v}{\Delta t} \;\rightarrow\; a\Delta t = \Delta v. \tag{7.3}$$

Using our expression for the change in momentum Δp and substituting,

$$\Delta p = m\Delta v = F_p \Delta t, \tag{7.4}$$

where we have used a form of Newton's 2nd Law to equate propulsive force F_p with the product of the total mass m and acceleration a. Equation (7.4) shows the relation between a change in momentum p, the propulsive force F_p, and time t. This product of force F_p over the time interval Δt is the *impulse* and is typically represented by the letter J,

$$J \equiv F_p \Delta t. \tag{7.5}$$

Think of impulse as *the input that leads to a change in the hull's momentum.* By exerting a propulsive force over a stroke's power phase, we change the hull's momentum, which means we change its speed.[38] This is apparent to anyone that has paddled a canoe, kayak, surfski, dragon boat, outrigger, SUP, or anything else. And the expression above for impulse will give us insight into what makes a hull go fast. Or go slow.

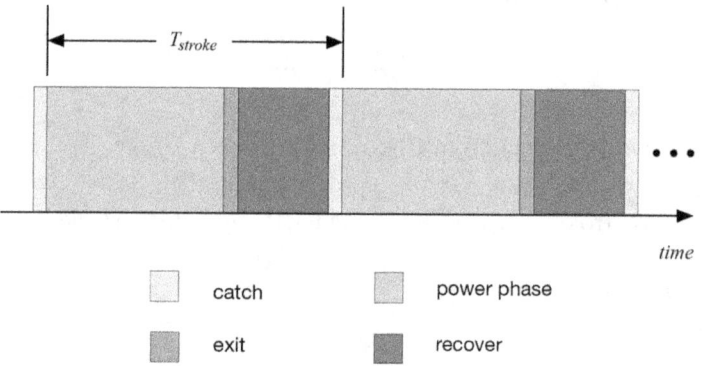

Figure 7.3: The "paddle cycle" elements and corresponding times.

First, let's define what we mean by this time interval Δt. Walking and running have *gait cycles* corresponding to leg movements during locomotion: lift, swing, step, and stance. Similarly, the *paddle cycle* comprises a sequence of events with corresponding time durations shown in Fig. 7.3.

We define the *power phase* as the event and corresponding time interval during which a propulsive force is exerted opposite to the direction of travel. The *exit* is the event and corresponding time interval when we draw the blade out of the water. The *recovery* is the

---

38 If we also change its mass, we've done something weird.

event and corresponding time interval over which the blade is out of the water and moving forward. And the *catch* is the event and corresponding time interval when we insert the blade into the water. Per our definition, the catch entails no force generation. A stroke's time duration is the sum of each element's time duration,

$$T_{stroke} = T_{power} + T_{exit} + T_{recovery} + T_{catch}. \tag{7.6}$$

Since Δt in the expression for impulse (7.4) corresponds to the time we exert a propulsive force, Δt is the duration of the power phase T_{power}. We'll see in a bit why this definition, while physically accurate, may be less useful to paddlers. First, we'll carefully define what we mean by propulsive force.

When we paddle, we exert forces on the paddle's shaft. In response, the paddle blade exerts a distributed force on the water: the pressure that is integrated over the blade's surface. A fundamental tenet of fluid mechanics is that pressure must act perpendicular to surfaces. Here, it acts perpendicular to the blade at every point on its surface, for all blade orientations.

The net force the blade exerts on the water is the sum of these distributed pressures. The resulting net force is always perpendicular to the blade, regardless of the blade angle. We'll represent the net force as the vector F, as shown in Fig. 7.4 for a straight shaft paddle and Fig. 7.5 for a bent shaft paddle.

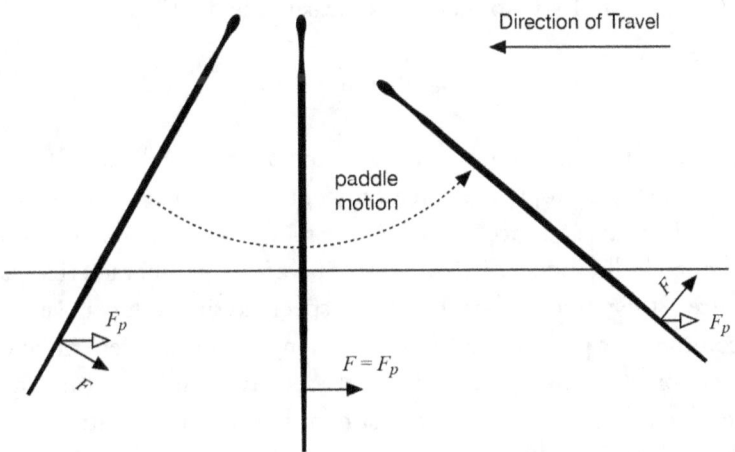

Figure 7.4: Paddle and propulsive forces, straight shaft, shaft angles –30 to +50 degrees.

We can infer the magnitude of the blade force F using (for example) strain sensors. And we measure it knowing force is always directed perpendicular to the blade's power face.

However, a hull is propelled (and slowed down) by forces acting coincident with the direction of travel. And the direction of travel is always parallel to the water's surface.[39] As shown in Figs. 7.4 and 7.5, the propulsive force acts opposite the direction of travel owing to

---

39 Unless you're trout scouting. As downriver paddlers say, "There are two types of paddlers: those who have swum in whitewater, and those who will."

Newton's 3rd Law of Motion. We see that the angle between the propulsive force and the blade's power face changes as the paddle rotates through the power phase.

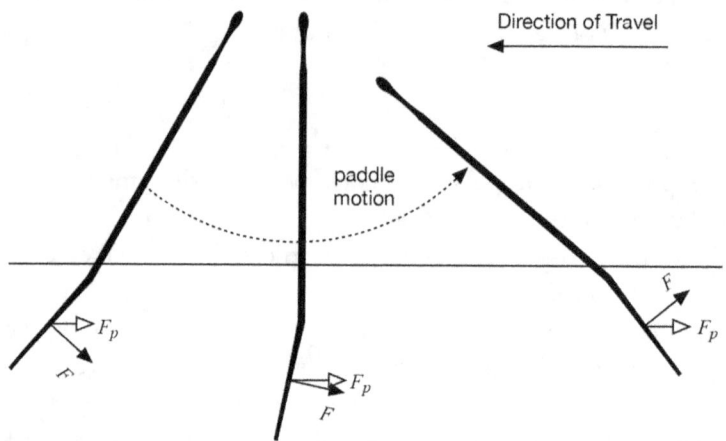

Figure 7.5: Paddle and propulsive forces, bent shaft, shaft angles –30 to +50 degrees.

Consequently, we must weigh the paddle's net force F as a function of the blade's angle to derive the propulsive force F_p. The way we do this is to measure the shaft angle and multiply the blade force by that angle's cosine, as we did in Chapter 6. We can (for example) use a shaft-mounted solid-state IMU to make this measurement. Thus,

$$F_p = F \cos(\theta - \phi). \tag{7.7}$$

where θ is the angle of the shaft from the vertical, and ϕ is the bend angle (if present) between the shaft and blade. The propulsive force only equals the blade force when the blade is perpendicular to the water's surface. Our impulse model is based upon the propulsive force.

The power phase's duration Δt is less useful if we wish to characterize paddling using *cadence*. Cadence is the number of strokes over a specified time interval, such as strokes per second (less common) or per minute (most common). We can infer stroke cadence using solid-state accelerometers mounted to the hull, detecting and counting the acceleration peaks due to each stroke. Since we can measure it, how can we use it?

Consider the example paddling propulsive force versus time curve plotted in Fig. 7.6. In this example, the cadence is 60 *spm* (strokes per minute) or one stroke per second. Consequently, the horizontal axis in Fig. 7.6 covers exactly one stroke. At zero seconds, the catch is complete, and the pull has started. The exit occurs at 0.7 seconds. Note that the paddle is still loaded – e.g., the propulsive force is greater than zero – at exit; we'll investigate that more a bit later. In our example, the peak force is 52.4 *lbf*. Over the power phase, the average force is 27.05 *lbf*, as indicated in the figure. This is the average force we'd use to compute impulse, with a corresponding time duration of T_{power}.

The average force over the entire stroke is useful because the stroke duration corresponds to the cadence's period. Per our definition, the propulsive force is zero outside the stroke's power phase. Consequently, the average propulsive force over the entire stroke is less than the average force over the power phase, as indicated in Fig. 7.6; we're now averaging over a longer period. For our example, the average propulsive force over the *entire* stroke is 18.94 *lbf.*

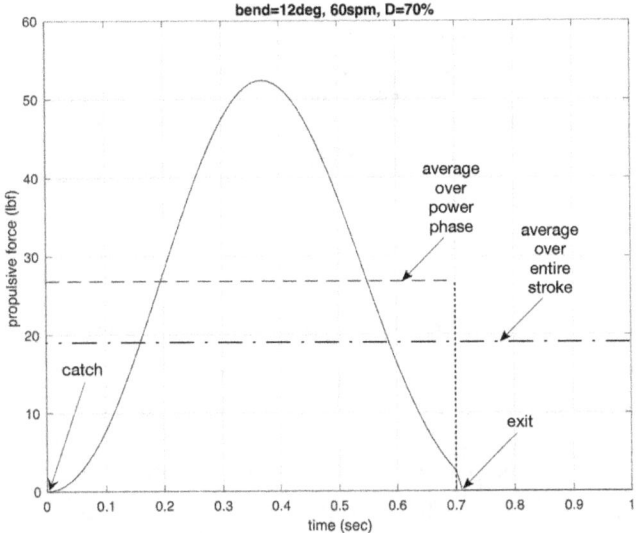

Figure 7.6: Time intervals, force, and force averages.

The force curve over time, and average force over the power phase, are quantities we can measure using an instrumented paddle. The average force is the integral over the power phase of the force curve, divided by the power phase's duration. Yeah, that sounds like calculus, and it is. But the integral is just the area under the force curve. We can compute the integral digitally by summing measured force sensor data.

Fortunately, the average force over the entire stroke is just the product of the average force over the power phase times a *duty factor D.* For any cyclic system, the *duty cycle* is the fraction of a period over which the system is active. Duty cycle describes systems that employ pulse width modulation (PWM), such as stepper motor controllers and particular radio (RF) and wired communications protocols. We'll parameterize paddling's duty cycle using a duty factor, the fraction of time over an entire stroke with a non-zero propulsive force. This is illustrated in Fig. 7.7. The paddling system is 'on' when the paddler exerts force through the paddle during the power phase, and it is 'off' the rest of the time. The duty factor D ranges between 0 (no propulsive force and thus no impulse at all) and 1 (zero exit, recovery, and catch time). Since no real-world paddling stroke has zero recovery time, we note that D may get close(r) to 1 but never equal it. You can also express the duty factor as a percentage.

The duty factor is related to % recovery time – the sum of the exit, recover, and catch times – by 100% minus D. A higher duty factor D means a faster recovery. For the example force profile in Fig. 7.6, D = 70%. If you multiply the average force over the power phase in Fig. 7.6 by 70%, you get the average force over the entire stroke. For our example, 27.05 *lbf* times 70% equals 18.94 *lbf.* Check.

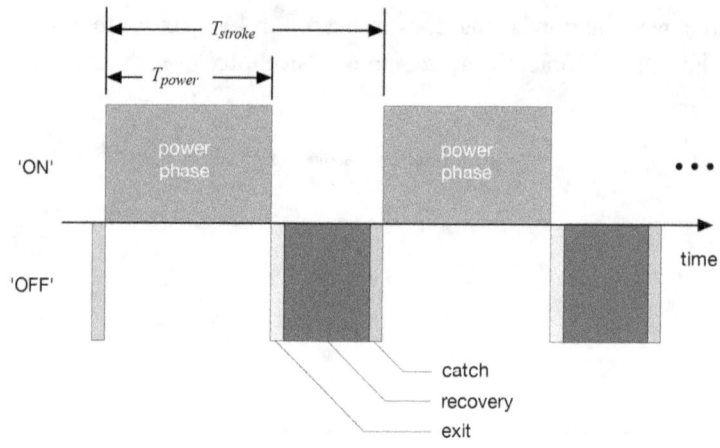

Figure 7.7: 'On' and 'Off' portions of the paddling cycle.

Examples

We'll model a bent shaft canoe paddle with properties and operating values listed in Table 7.1. The analysis applies to any paddling discipline merely by specifying other values.

TABLE 7.1: BASELINE MODEL PARAMETERS

ρ = 1000 kg/m^3 (water density)

Blade area = 0.1 m^2

C_{d0} = 1.15 (blade drag coefficient)

Stroke duration = 1 sec

Power Phase duration = 0.7 sec

Peak force scaling[40] = 1

Using the model developed by Caplan this paddling scenario generates the propulsive force versus time curve in Fig. 7.8. The paddle moves over a shaft angle range from −30 degrees to +50 degrees. We covered this model in detail in Chapter 6.

Fig. 7.8 shows three force curves. Each has the same peak force, 52.4 *lbf*. The solid line is the force curve for our baseline paddle, whose parameters are listed in Table 1. The dashed line is the force curve where the power phase is 60% (D = 60%) of the stroke period for a 60*spm* cadence (T_{stroke} = 1 sec). The dash-dot line is the force curve where the power phase is 50% (D = 50%) of the stroke period for a 60*spm* cadence (T_{stroke} = 1 sec). In other words, we move from a 30% recovery fraction to a 50% recovery fraction of the stroke duration, with quicker force production but slower recovery.

We compare the impulse generated by each force curve by computing the average impulse over the power phase. Impulse has units of force times time; here, we'll use MKS (meter, kilogram, second) units. The results are listed in Table 7.2.

---

40 This scaling is used in our MATLAB model to scale the peak force. When equal to one, the resulting peak force in our example is about 52.4 lbf, or 233 Newtons.

TABLE 7.2
D = 70%, impulse = 84.22 *N-sec*
D = 60%, impulse = 72.19 *N-sec*
D = 50%, impulse = 60.16 *N-sec*

Table 7.2 shows that for a fixed cadence, shortening the power phase while maintaining the same peak propulsive force and force-time curve results in lower impulse. Consequently, for a given hull you go slower with a slower recovery.

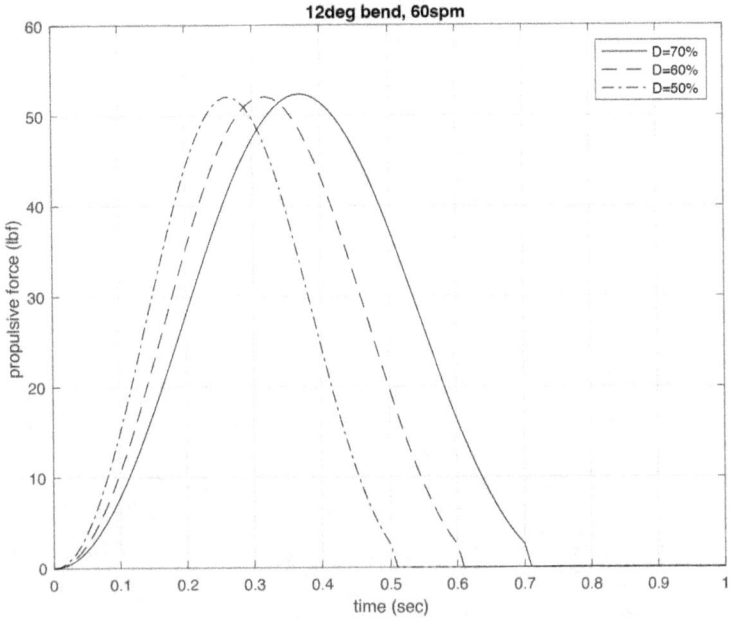

Figure 7.8: Equal peak force curves with different duty factors.

How can we overcome this limitation? We increase the cadence so that the type of stroke depicted in Fig. 7.8 provides the same *cumulative* impulse over a *series* of strokes, such as over a minute or more. In other words, if your stroke generates less impulse – less area under the force curve – you need to up the stroke rate. While this isn't news to most paddlers, the analysis tells us *why* it's so. Our model lets us compute how much the cadence must increase for equal propulsion. For our D = 60% stroke force curve, cadence must increase by 16.7%, corresponding to 70*spm* or a 0.86-*sec* stroke duration. For our D = 50% stroke force curve, cadence must increase by 41.4%, corresponding to 84*spm* or a 0.71-*sec* stroke duration. Each lead to a duty factor of 70% for a shorter stroke duration/faster cadence. We've essentially reproduced our baseline force/stroke duration curve in miniature, just with more repetitions over time.

Alternately, one can keep the force curve shapes, and their respective duty factors defined in Fig. 7.6 and scale the peak propulsive forces. These new curves are plotted in Fig. 7.9.[41] By scaling the peak forces, each stroke's impulse is equal for the same cadence – 60 *spm*.

---

41 The notation "equal areas" at the top of this figure indicates equal areas under the force curves.

Consequently, each stroke will equally increase the hull's momentum and thus its speed. The peak forces for the three curves are 52.4 *lbf*, 61.1 *lbf*, and 73.3 *lbf*, respectively.[42] So if you want a slower recovery for a fixed cadence, you need to up each stroke's peak force for equal propulsion.

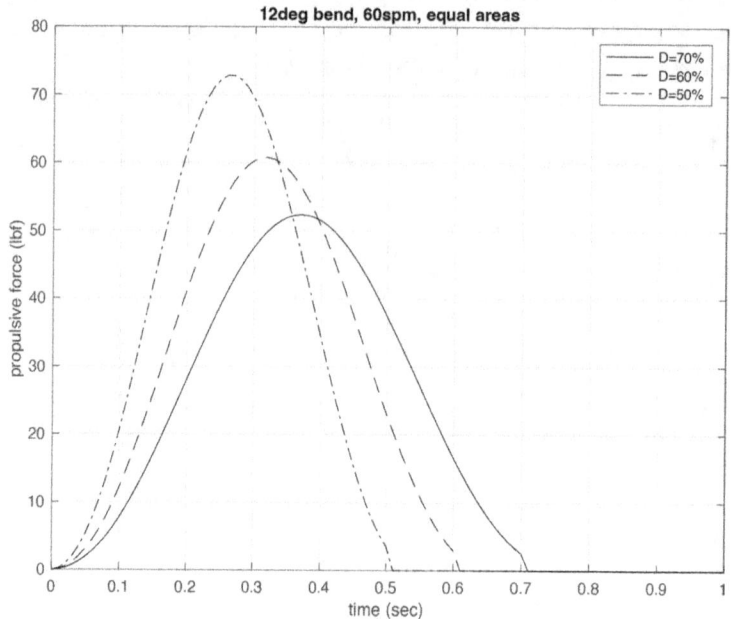

Figure 7.9: Varying peak force curves with different duty factors and equal impulse.

Finally, what happens if we cut a stroke short(er)? Our baseline power phase ends with the paddle blade still slightly loaded, as shown in Figs. 7.6, 7.8, and 7.9. What happens if we shorten the power phase duration of this stroke – decrease the duty factor D – for a fixed cadence? This is shown conceptually in Fig. 7.10. The force curve is generated using the parameters from Table 1. The baseline entails an exit at 0.7 *sec*. The figure also shows force curves with exits at 0.6 and 0.5 *sec*. How much do we impact the stroke's ability to propel a given hull with earlier exits?

We'll once again compare these scenarios using impulse. The results appear in Table 7.3.

TABLE 7.3

Exit at 0.7 *sec*, impulse = 84.22 *N-sec* (100%)

Exit at 0.6 *sec*, impulse = 80.46 *N-sec* (95.5%)

Exit at 0.5 *sec*, impulse = 68.61 *N-sec* (81.4%)

We see from Table 7.1 that exiting at 0.6 *sec* only decreases impulse by 4.5%. To regain the "lost" impulse, we can up the cadence from 60*spm* to ~62.8*spm*.

Personally, I prefer to exit when the blade is still a bit loaded. By shortening the power phase about 14% in our example, we haven't lost much area under the force curve. We've

---

42 These scaling factors are just the ratios of the duty factors.

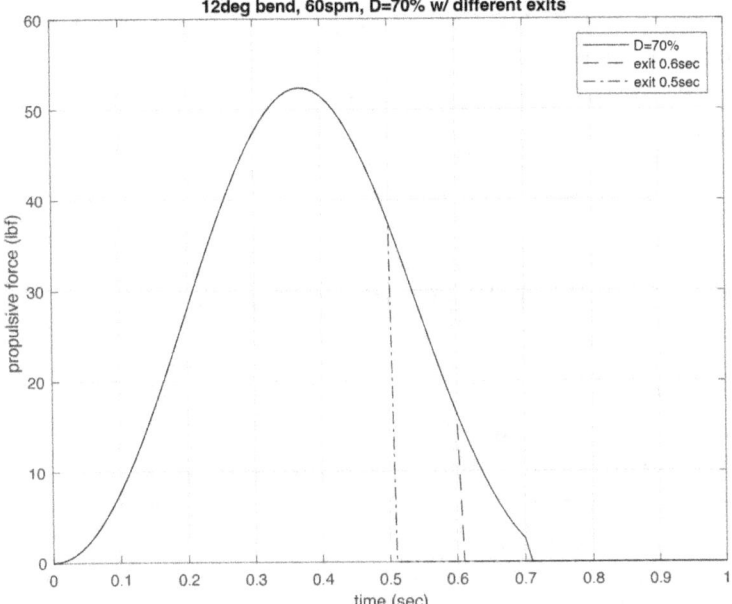

Figure 7.10: Baseline force curve with different exits.

only cut off a bit more of the force curve's "tail," as seen in Fig. 7.10. The sole criteria for determining and comparing impulse is changes in the area under the force-time curve over the stroke. And the increase in cadence for equal impulse is small for this case.

Things are different when we exit at 0.5 *sec*. Impulse decreases by 18.6% compared to our baseline, requiring we up the cadence from 60*spm* to ~71*spm* for equal impulse. In this case, the blade will be strongly loaded at the exit. This scenario may be a bit extreme, if not unrealistic. But an earlier exit entails cutting off some of the area under the force-time curve. The earlier the exit, the more impulse we lose unless we increase cadence.

How Do I Use This Information?

There are several ways to increase impulse considering the examples above. You can increase stroke cadence to increase total impulse over a series of strokes. However, cadence by itself is a factor but not the sole metric defining propulsive input. Also, when increasing cadence, you need to maintain the same average impulse and duty factor. Otherwise, you're not increasing impulse.

You can increase impulse by increasing the stroke duty factor, i.e., the percentage of time you're exerting a paddling force to move the hull forward for a given cadence. This shortens your total recovery time as a fraction of the stroke time.

Referring to Fig. 7.7, we note that the total recovery time is the sum of the exit, recovery, and catch times. None of these contribute to propulsion; they just set the paddler up for the next stroke. This begs the question of whether a particular style of exit, recovery, or paddle blade entry makes you go faster. Yes, they can, *but only if you minimize your time not exerting a paddle force.* The moral of this story is to initiate a quick, deep catch so that you can start pulling and contribute to the area under the force curve. Whatever technique accomplishes this *for you* is best; physics doesn't pick sides. Then, exit when there isn't a lot of felt pressure on the blade since you're no longer contributing to the area under the force curve. A force curve with a long, low-amplitude tail also doesn't shorten recovery time.

And finally, we can increase impulse by increasing the product of the propulsive force over a stroke times the power phase duration. In Chapter 6, we learned to represent the time-varying propulsive force via

$$F_p = \frac{1}{2}\rho A C_{d0}\cos(\theta - \phi)\left(v_{hull} - v_{paddle}\right)^2, \tag{7.8}$$

where ρ is the water density, A is the paddle blade area, C_{d0} is the blade drag coefficient when the power face is perpendicular to the water surface, θ is the time-varying shaft angle, ϕ is the paddle bend angle with respect to the shaft, and v_{hull} and v_{paddle} are the hull and paddle speeds in the direction of motion, respectively, which are also time-varying functions. Impulse equals the area under this curve.

We learned in Chapter 6 that we generate the greatest peak force if the maximum velocity difference occurs when the blade face is perpendicular to the water's surface. This angle maximizes the cosine function. Both the cosine and the square of the relative velocity are positive functions, so their product will be positive, too.[43] This "phase alignment" of velocity and blade angle is a matter of technique and paddle fitting. Increasing the peak propulsive force may increase impulse subject to the constraints laid out above.

Another way to increase impulse, assuming all other factors remain the same and are independent of these changes, is to increase the blade area A and the blade normal drag coefficient C_{d0}; there's not much you can do about the water density ρ. After a certain point, however, the blade will become unreasonably large[44] and too tiring to use except for the most muscle-bound among us.

Equation (7.8) shows that you can increase the peak force by increasing the relative velocity between the hull and paddle speeds. This is like the advice given to rowers: focus on handle speed once the oars have caught.

These three factors – impulse, duty factor, and cadence – are inextricably linked. What performs these functions are people. Paddlers have different physiologies, strengths, techniques, and motivations. None of us have separate dials for cadence, paddling duty factor, and impulse. Nonetheless, the analysis points to focuses for tuning technique: A quick, deep catch to minimize T_{catch}; maximizing the hull vs. paddle speed when the blade is around and at the vertical; exiting when you're not contributing a lot of propulsive force; reducing in-air recovery time using whatever technique works for you.

---

43 This is another way of saying that the propulsive force is always greater than or equal to zero. Common sense here is more concise than the math.

44 Some of you may remember the old "banjo" paddles of the 1970s and 1980s.

Take-Aways

- Momentum and changes in momentum characterize hull speed. Impulse, the product of the propulsive force generated by our paddles and the duration of the stroke's power phase, drives changes in momentum.

- Impulse provides a quantitative way to compare the impact of stroke dynamics – propulsive force versus time, peak force, stroke cadence, and stroke duty factor – on propulsion. While we may consider each element separately to develop paddling and training focuses, it is the combination of impulse, duty factor, and cadence (as well as the hull's total drag) that ultimately determines hull speed.

- Shortening the stroke's power phase reduces impulse for a given cadence and peak propulsive force. You can recover the "lost" impulse by increasing stroke cadence by the inverse of how much the stroke is shortened.

- If you decrease the power phase duration for a given cadence and propulsive force curve, you must increase the peak propulsive force to generate the same impulse.

- For a given propulsive force curve, you can use impulse to assess the impact of shortening the power phase via an earlier exit. If the area under the "tail" of the force curve is small compared to its entire area, the impact of shortening the stroke may be negligible. This corresponds to exiting the stroke while the blade is still loaded, which may set the paddler up for a quicker recovery.

- Distance per stroke is a secondary (derived) parameter for characterizing propulsion. Cadence and impulse are the primary factors to focus on if at all possible.

Extra Credit: Distance per Stroke and Cadence

The impulse delivered over an entire stroke is defined precisely by

$$\Delta p = \int_{t_{catch}}^{t_{exit}} F_p(t)\, dt \, \cdot \tag{7.9}$$

The integral limits correspond to the end of the catch and the start of the exit. This equation is just a concise way of expressing what took us a few paragraphs to explain above. Yay, calculus!

As an aside, we know that paddlers come in all sizes and shapes. It can be instructive to normalize the impulse by a paddler's weight to calculate a non-dimensional metric, often referred to as *specific impulse*, which accounts for differences in paddler sizes that can correlate with differences in height and muscle mass.

· · ·

Next, we'll examine the role of distance per stroke d via the relation between distance per stroke, propulsive force F_p, and work W. Work is defined as the product of force and distance traveled. In terms of the propulsive force, for each stroke

$$W = F_p d \quad \rightarrow \quad F_p = \frac{W}{d} \, . \tag{7.10}$$

Consequently, we can write our definition for impulse Δp in terms of work W as

$$\Delta p = \frac{W}{d} \Delta t \, . \tag{7.11}$$

We can rewrite this in a (soon to be) convenient form as

$$\Delta p d \frac{1}{\Delta t} = W \, . \tag{7.12}$$

Work is also the time integral of paddling power P. We'll use the *average* power over a stroke as we did for our propulsive force. As a result, the integral becomes the product of average power P and the stroke duration,

$$W = P \Delta t \, . \tag{7.13}$$

As we'll see in Chapter 12, when cruising in steady-state average power is related to average velocity v via the hull's dimensional drag coefficient C_D,

$$P = C_D v^3 \, . \tag{7.14}$$

Combining these two expressions yields

$$W = C_D v^3 \Delta t \, . \tag{7.15}$$

Substituting this into our "convenient" result above,

$$\Delta p d \frac{1}{\Delta t} = C_D v^3 \Delta t = C_D v^2 d \, , \tag{7.16}$$

where we have used the fact that the velocity over a stroke is equal to the distance per stroke d divided by stroke's duration Δt. Notice how the distance per stroke appears on both sides of this equation. Consequently, we can divide both sides of the equation by d and solve for the average velocity over a stroke:

$$v = \sqrt{\frac{\Delta p}{C_D} \frac{1}{\Delta t}} \, . \tag{7.17}$$

The average velocity in Equation (7.17) does *not* depend on the distance per stroke; it does not appear in the equation. This tells us that distance per stroke *results from* applying impulse through our paddling strokes; it is not a cause. Distance does not produce speed. Indeed, we can express average speed using distance per stroke; that's high school physics. But the distance traveled with each stroke depends on how much "oomph" we apply to achieve a velocity. It also depends on the hull's hydrodynamics. A sleek, fast hull runs longer per stroke than a less sleek, slower hull.

The expression for average velocity depends on the square root of the impulse Δp. It depends inversely on the square root of the hull's drag coefficient and the stroke's duration. The drag coefficient "represents" the hull. A less sleek hull has a larger drag coefficient, and the equation shows it will have a slower average speed. A sleeker hull has a smaller drag coefficient, and the equation shows it will have a faster average speed. This reinforces why distance per stroke is a result, not a cause.

As a check on our result, using our expression for Δp in terms of propulsive force and stroke duration shows

$$ v = \sqrt{\frac{F_p}{C_D}} \quad \rightarrow \quad C_D v^2 = F_p. \tag{7.18} $$

In steady-state, the propulsive force balances the drag force, which is proportional to the velocity squared times the drag coefficient. (Note that we'll derive this in Chapters 11 and 12.) Check!

Further Reading

Paul Gaudet, Thomas Blackadar, and Steven Oliver, "Measuring foot contact time and foot loft time of a person in locomotion," U.S. Patent Number 6,018,705, January 5th, 2000

Kenneth P. Clark and Peter G. Weyand, "Are running speeds maximized with simple-spring stance mechanics?", *Journal of Applied Physiology* **117**, pp. 604 – 615 (2014).

Nicholas Caplan, "The Influence of Paddle Orientation on Boat Velocity in Canoeing," *International Journal of Sports Science and Engineering* **3**(3), pp. 131-139 (2009).

Interlude

INTERLUDE

Tales of Power

Preview

In this interlude between foundational chapters on science, and chapters on application, we'll look at how things don't always work out with our analyses. Blind spots can lead us on long and fascinating trips to nowhere – like believing I've elegantly solved the paddling power measurement problem. (Spoiler alert: I haven't.)

We'll re-introduce Newton's Laws of Motion in their original wording. The insights embodied there counter a habitual tendency to start by writing down equations. We'll also carefully consider frames of reference, keeping our wits about us as we move between Earth-centric and paddler-centric views.

Introduction

You don't often see articles about why or when things don't work out in the science and engineering literature. To the uninitiated, every press release or publication teems with breathless tales of world-changing achievement. But they can be a Potemkin reality presenting 100% inspiration and success with 0% perspiration, without any hint of dead ends or – eek! – failure.

Now bolts out of the blue sometimes occur, leading to surprising and even significant results. I drafted my master's thesis in a single weekend while I was ostensibly visiting my girlfriend. An intractable mess of equations – and months of dead ends – untangled itself on the drive to visit her. But those moments are rare and precious – as are weekends with your girlfriend.

Instead, actual science and engineering R&D is often 10% inspiration and 90% perspiration. A lot of dead ends entail work that produces no usable result. The hope is that no matter what, you learn something valuable along the way, which prepares you for the next foray.

In case you're wondering, yes, there are a lot of hen scratchings, simulations, and data crunching that end up on the virtual cutting room floor while "doing engineering." What sees the light of day is the work that survives revision and scrutiny.

In this interlude I invite you to look at one of my adventures down the rabbit hole (unsuccessfully) pursuing a simple way to measure paddling force and paddling power. Before we start, note that what I initially proposed *won't work*. Physics proves that, in general, the approach can't. We'll start by making some flawed assumptions.

What Is Force?

Consider the solo paddler depicted in Fig. I.1. As they paddle, they exert a force through the blade via the shaft, directed toward the stern.

Newton's 3rd Law of Motion states:

> To any action there is always an opposite and equal reaction; in other words, the actions of two bodies upon each other are always equal and always opposite in direction.

Here, "action" means force. Since a paddle force acts in one direction, there must be an equal force (the reaction) in the opposite direction of the paddle force. We'll call this the propulsive force $F_p(t)$.

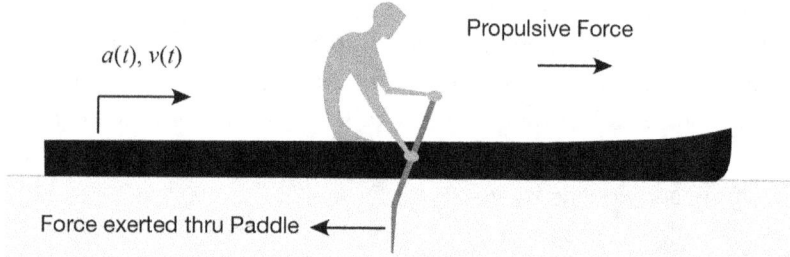

Figure I.1: Paddler and forces.

Newton's 1st Law of Motion states:

> Every body perseveres in its state of being at rest or of moving uniformly straight forward, except insofar as it is compelled to change its state by forces impressed.

In other words, the hull is stationary until you exert a paddle force, and "compel" the canoe to "change its state by force impressed." Further, if the hull is either stationary or already moving at a steady speed – its "state… of moving" – you must apply a suitable propulsive force to increase speed. The canoe's state is represented, for example, by a velocity $v(t)$ in the same direction as the propulsive force.[45]

Newton's 2nd Law of Motion states:

> A change in motion is proportional to the motive force impressed and takes place along the straight line in which that force is impressed.

This implies that changes in hull velocity are proportional to the propulsive force, and these changes occur in the same direction as that force.

All three Laws of Motion appeared in Newton's *Mathematical Principals of Natural Philosophy*, published in 1687. To make the 2nd Law more tangible, you must dig into the text. The "alteration of motion" recited in the Law is the change of an object's momentum. Recall from Chapter 4 that momentum is the product of mass and velocity. Since our hull and paddler have constant mass, this equation takes on the familiar form

$$F_p(t) = ma(t), \tag{I.1}$$

where $a(t)$ is the hull's acceleration. Acceleration is the change in velocity over time.

This equation has a beguiling simplicity. But to measure the paddle force directly, you must place sensors on the paddle, along with signal conditioning electronics and a power source. You'll most likely need a radio transmitter to send the sensor signals to either a data recorder or a smartphone. But Newton's 3rd Law states that this force must equal the propulsive force – note that this is my first mistake. Equation (I.1) implies that if I know the combined mass of the hull and paddler, then measuring the acceleration of the hull in the direction of travel lets me calculate the propulsive force. This then equals the paddle force! Woo hoo!

Note that this inference, based upon mistaken prior assumptions, is wrong. But bear with me.

As we saw in Chapter 6, the paddle force curve resembles a portion of a time-shifted cosine function, as shown in Fig. I.2. Since the system's mass is constant, we can assume that the propulsive force will have the same general character. Some apps measure acceleration using the accelerometer built into most modern smartphones.[46] They record this data to a file for later analysis. So all I had to do was download an app to my smartphone, secure the phone in the bottom of my hull, and record data. It sounded almost too easy! Fortunately, it was early March, and the rivers near me were still frozen. But I get ahead of myself. First, we need to link force with paddling power.

---

45 For those of you with a background in dynamic systems theory, velocity is a state variable; it represents a state of the system.

46 In smartphones, solid-state MEMS accelerometers are often used to determine the phone's spatial orientation.

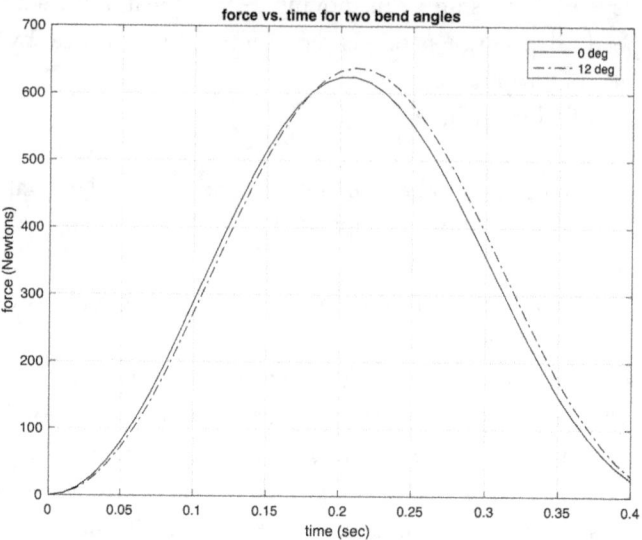

Figure I.2: Normalized paddle force vs. time.

Tales of Power, Part 1

If you can compute paddling force and access the hull's velocity data, you can calculate paddling power. As we learned in Chapter 1, power is the product of force and velocity. We already have access to acceleration data from the smartphone's internal sensor. We also know that acceleration is the time derivative of velocity. Consequently, you can employ differentiation's inverse operation, known in intro calculus courses as an "anti-derivative" or more commonly as an indefinite integral, to compute velocity. Power is then expressed – incorrectly, but again bear with me – as

$$P(t) = ma(t)v(t),\tag{I.2}$$

where v is the computed velocity. Looks great. Drop an accelerometer in the bottom of your hull, and voila! With a bit of computing, paddle force and power are yours for the taking.

The problem is, it's all wrong.

What's missing? Recall from Chapter 2 that there is a third force acting here: the drag force. Newton's 1st Law of Motion requires that you incorporate *all* forces acting on an object, *not just the ones you prefer*; we'll return to this point in a bit. This means Newton's 2nd Law must reflect all forces on the hull:

$$F_p(t) = ma(t) + F_{drag}.\tag{I.3}$$

The drag force acts in opposition to our motion. That's why it's on the opposite side of the equation from the propulsive force. Drag is also a property of the hull and its hydrodynamics. So, to calculate the propulsive force, you must also measure the drag force. Making flawed assumptions comes back to bite you once you correctly apply the laws of physics. It's embarrassing. At least I came to my senses.

Tales of Power, Part 2

It was time to go back to the drawing board, or at least to my notebook. Will a "new" analysis survive the impartial yet exacting scrutiny of physics?

Consider the solo paddler depicted in Fig. I.3. We now have (hopefully) identified all the forces on the hull. A careful reading of Newton's 3rd Law of Motion indicates, "To *every* action there is always opposed an equal reaction" [emphasis added]. Every means, well, *every*, not just the actions (e.g., forces) that are convenient.

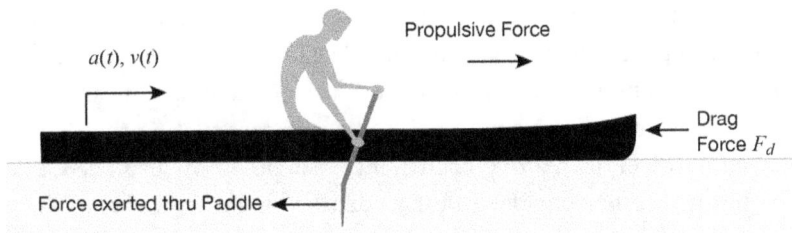

Figure I.3: Paddler with more forces.

From Chapter 2, we note the drag force F_{drag} is proportional to the square of the hull's velocity,

$$F_{drag} = C_D v^2(t),\tag{I.4}$$

where C_D is the hull's dimensional drag coefficient. Over the speed range of a paddled hull, the drag coefficient is a constant and can be measured via field tests or estimated via numerical simulation. So, we can rewrite Equation (I.3) as

$$F_p(t) = ma(t) + C_D v^2(t).\tag{I.5}$$

Power P, which is the product of force and velocity, can then be expressed as

$$P(t) = ma(t)v(t) + C_D v^3(t).\tag{I.6}$$

Eureka! Paddling power expressed using (1) known or readily knowable quantities (combined mass of hull plus paddlers and drag coefficient), (2) a quantity that is measurable in real-time (acceleration), and (3) another that is computable from measurements (velocity).

Or so I thought. Something was niggling at me and wouldn't stop. Can you see why this result, in general, is not correct? (Hint 1: Total forces. All of them. Hint 2: Frames of reference.)

We've once again neglected all the forces that can act on a hull. Let's see why with an update of our paddler and force diagram, shown in Fig. I.4.

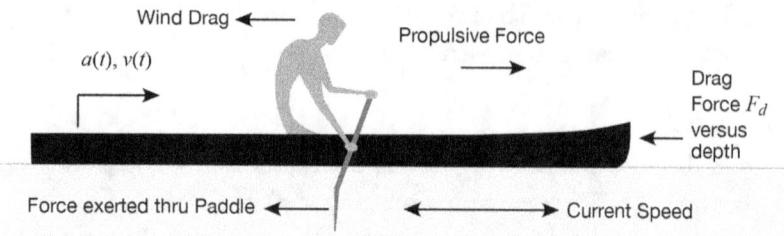

Figure I.4: Paddler with even more forces and effects.

Along with the various forces and state variables, Fig. I.4 includes the action of wind, be it a headwind, tailwind, or something in between. Wind exerts a force that can be helpful or not – and can mess with things if it kicks up big waves. Next, as we learned in Chapter 3, there are depth-dependent effects that can impede us. These act as another drag force in shallow water. The combination of wind and depth-dependent drag means that we haven't accounted for "every action," as Newton pointed out in formulating the Laws of Motion. So now we need a model of the drag coefficient vs. depth. We also need a depth meter to tell us which drag coefficient value to use as depth changes over time. Plus, we'll need something to measure the wind and its direction, along with a 3-D model of wind drag for the hull and paddler. The problem has become complex.

Next, consider current. You might ask, why do I care? Let's say you're paddling in deep water with no wind. Can't we use the model derived above to infer paddling force and power from the acceleration measurement?

Accelerometers are sensors that measure acceleration in an *inertial frame of reference*. This means accelerometers measure acceleration with respect to the Earth, which is considered stationary.[47] The in-hull acceleration measurements described above are referenced to the shoreline. Why might this be a problem? Consider the following scenario: Paddling upstream in a current that exactly equals your paddling speed. In this case, you're stationary with respect to the shore. As you paddle, there will be some acceleration and deceleration in the shore frame of reference. But the velocity you compute by integrating the accelerometer signals will be the increments of added speed during the stroke's power phase, not your speed with respect to the current. Remember that the drag force depends upon *the hull's movement relative to the water, not the shoreline*. Oops. Wrong velocity.

Can you fix this reference frame issue by using a GPS receiver to calculate your locally average velocity over ground – i.e., tied to the inertial frame of reference? Could we use the accelerometer signal to compute velocity increments above this average and deduce what's happening during the power phase? And perhaps drop current meters all along our course? Well, it's possible. But doing so adds even more complication to what I had hoped might be a single, simple measurement. And it doesn't address the role of wind forces in the "every action" force constraint.

Well, at least it was a fun way to get "No" for an answer.

---

47 Or, more precisely, not accelerating.

Lessons Learned

In their landmark paper, "Sound waves in rooms," Philip Morse and Richard Bolt wrote of a need to "salt our analysis with liberal doses of common sense." And that's certainly the case here. It's very, very easy to become enamored with our tools. With a narrow focus we lose sight of what we're trying to accomplish in the real world. It's challenging to let go of an idea, even when its range of useful applications becomes vanishingly small.

As I look at the wall of engineering, mathematics, and science texts in front of me right now, I'm amazed that it's possible to keep enough of that stuff in mind to accomplish anything; to make anything work. A fellow engineer once asked if I was worried that a complex feedback control system I designed might have a sign error buried in its parameters. My reply (which I have since deeply regretted) was, "We'll find out when we turn it on." He turned to face me, looked me in the eye, and noted, "Remind me not to fly in any spacecraft you design." So, we simulated and tested the daylights out of that control system before integrating it into the final hardware. Going forward, I do the best that I can to check my work along the way, asking over and over, "Are you sure?"

I hope you've come to appreciate the value of salting our analysis with liberal doses of common sense. All factors must be accounted for to develop a meaningful (or even workable) engineering system. If you choose to neglect certain factors, only do so after they are fully understood. Then assess their relevance and impact. You know, solid engineering practice 'n stuff. The good news regarding the analysis presented here – at least for me – is I knew when to put fingers to the keyboard and when to not. Not all eureka moments are actually eureka moments.

Further Reading

P.M. Morse and R.H. Bolt, "Sound waves in rooms," *Reviews of Modern Physics* (**16**), pp. 69-150 (1944).

Kenneth Fyfe, James Rooney, and Kipling Fyfe, "Motion Analysis System," U.S. Patent 6,513,381 (February 4th, 2003).

94

Part II

Applications

CHAPTER 8

There and Back

So, which is faster: paddling an out-and-back course without current or with the current? You lose time going upstream into current, but you regain it going downstream, don't you? Shouldn't a round trip should take the same amount of time? Well, sometimes intuition leads us to conclusions not supported by just a little algebra.

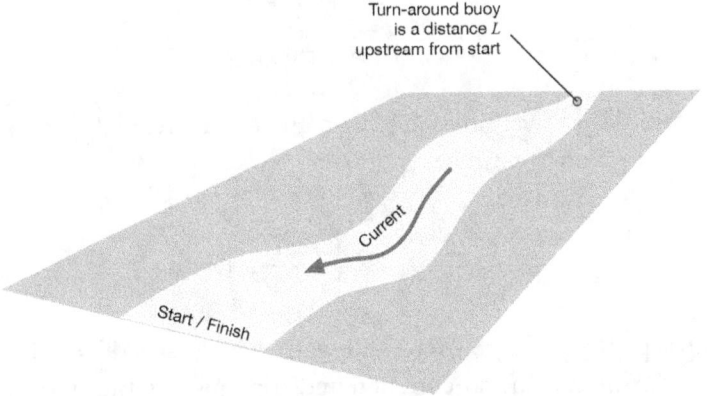

Turn-around buoy
is a distance L
upstream from start

Current

Start / Finish

Figure 8.1: Out and back course concept.

Consider an out-and-back course that first travels upriver, then back downriver to the starting point as depicted in Fig. 8.1. If the one-way distance to the turnaround buoy is L, then the round-trip distance is twice this, $2L$. When there is no current, and you paddle this course at an average speed V (and, unlike me, could magically make the buoy turn in zero time), then your round-trip paddling time T_{still} equals

$$T_{still} = \frac{2L}{V}. \tag{8.1}$$

For simplicity, we'll assume that the river has a uniform current along its entire length, with average speed U. This means when you travel upstream, your speed equals your still water speed minus the current's speed, $V - U$. In other words, the current slows you down. As a result, your upstream time to the turnaround buoy T_{up} is

$$T_{up} = \frac{L}{V - U}. \tag{8.2}$$

As you can see, since your upstream speed $V - U$ is less than your still water speed V, your time to the turnaround buoy T_{up} is now greater than your time to cover this distance in still water due to the river's current.

With current, your downstream speed now equals the sum of your quiet water paddling speed V plus the speed of the current, $V + U$. As a result, your time from the turnaround buoy back to the start, T_{down}, equals

$$T_{down} = \frac{L}{V + U}. \tag{8.3}$$

Since your downstream speed $V + U$ is larger than your still water speed V your downstream time will be faster than if there was no current. Paddlers everywhere are familiar with losing time going upstream and gaining time going downstream.

Your round-trip time is the sum of your upstream time T_{up} and your downstream time T_{down},

$$T_{current} = T_{up} + T_{down}. \tag{8.4}$$

Using our expressions for travel times upstream and downstream T_{up} and T_{down}, after a bit of rearranging, your round-trip time is

$$T_{current} = \frac{2L}{V \cdot \left(1 - \dfrac{U^2}{V^2}\right)}. \tag{8.5}$$

This relationship is plotted in Fig. 8.2 for a still-water average speed $V = 6.4$ *mph*, and round-trip distance $2L = 4.95$ *miles*. The no-current round-trip finishing time is 0:46:24. If the river had no current, $U = 0$ and $T_{current} = T_{still}$; check. However, if the current is non-zero, then $T_{current}$ will always be greater than T_{still} since the denominator in equation (8.5) will always be less than V – the term in parenthesis in the denominator is always less than one. As a result, on an out-and-back course that travels up and down a route with current, you can't recover the time lost going upstream on the faster downstream leg. The time relationship is not linear with speed, both for each leg and the total time.

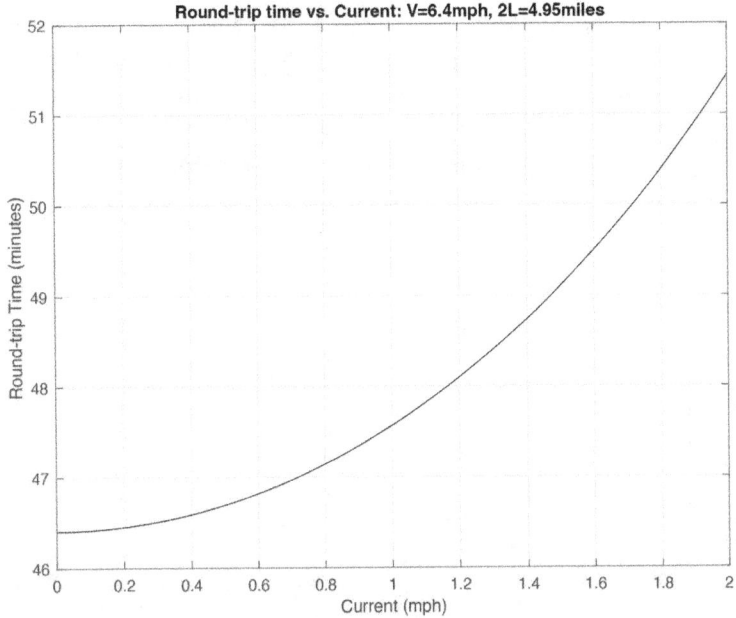

Figure 8.2: Round-trip time vs. current speed for a fixed paddling speed.

If the current is much slower than your still water paddling speed, $U \ll V$, the term U^2/V^2 in the denominator of equation (8.5) is much smaller than 1. In that case, you can approximate equation (8.5) as[48]

$$T_{current} \sim \frac{2L}{V} \cdot \left(1 + \frac{U^2}{V^2}\right) \quad \text{for} \quad U \ll V. \tag{8.6}$$

Now the term $2L/V$ is your still water travel time. For non-zero current speed U, the term U^2/V^2 is greater than zero, which means $1 + U^2/V^2$ is greater than one. So your still water travel time will be multiplied by a number that is greater than one, and your round-trip time will always be slower in the presence of current, *no matter what*. Going faster on the downstream run because of current will not make up for going slower on the upstream run. Why? Because while you travel faster downstream, you spend less time going faster than the amount of time you travel more slowly upstream. The longer time traveling more slowly gets you.

There are two other interesting limiting cases in equation (8.5). First, if the current is running exactly as fast as you can paddle, e.g., $U = V$, then your round-trip travel time becomes infinite; you never get to the turnaround buoy or anywhere else for that matter, as you're just sitting still. The denominator of equation (8.5) equals zero when $U = V$, hence

---

48 For those of you keeping score at home, this approximation is derived using a binomial expansion.

the infinite travel time. Infinite travel time is how mathematicians say, "You can't get there from here."

Second, if the current is going faster than you can paddle, i.e., $U > V$, then your upstream travel time T_{up} expressed in equation (8.2) becomes negative – you're going backward, but not back in time! (Don't worry; we haven't invented a time machine.) The result is just an artifact of the algebra, a precise, mathematical way of stating that you're going backward.

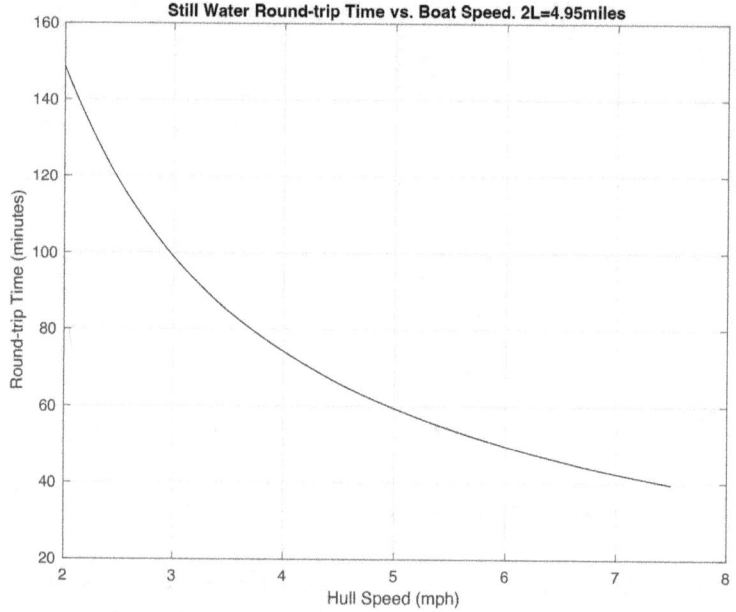

Figure 8.3: Round-trip time vs. hull speed, no current.

So, is the fact that you "lose" round-trip time when there is current all that surprising? Not really, if you change your perspective. We habitually think of the round-trip transit problem in terms of distance or speed. For a fixed time, if we double the speed, we go twice as far. But for a fixed *distance* course, time and speed have an inverse relationship. As a result, even when there is no current, the relation between round-trip course time and boat speed is not a simple linear proportionality, as shown in Fig. 8.3.

This curve "flattens out" at higher hull speeds, reflecting a more linear proportionality the faster you go. More generally, in the presence of current, a faster hull speed reduces the effect of current on round-trip course time but does not eliminate it, as indicated by equation (8.6).

The only way to make up the time deficit is to paddle faster than the still water speed on one of the legs. If you're racing, you might pick up the pace over the last downstream mile or so. But, alas, paddling faster on one of the legs violates an assumption of this analysis; we assumed that your still water paddling speed V is *constant*. We must always keep the underlying assumptions in mind when considering the conclusions of any analysis.

Take-Aways

- For a fixed distance route, time and speed have an inverse relationship.

- Even when there is no current, the relation between round-trip course time and boat speed is not a simple linear proportionality.

- Going faster on the downstream portion of a round-trip course does not make up for going slower on the upstream portion. While you travel faster downstream, you spend less time going faster than the amount of time you travel more slowly upstream. The longer time traveling more slowly gets you.

- If you must choose between paddling harder on the upstream or downstream portion of an out-and-back course, pick the upstream leg. This is because you'll spend more time paddling upstream and you can chip away at the time difference we derived above. All of this goes out the window, of course, if the river banks are highly crenelated, the depth varies significantly cross-stream, and the river is full of rocks and downed trees.

Afterword

While writing this chapter, I recalled a problem noted at the 2016 Rio Olympics. The swimming pool may have had some current in the high numbered lanes. (See, for example, https://www.theguardian.com/sport/2016/aug/18/olympic-pool-current-swimming-results-rio-2016.) Swimming events are timed no more precisely than 0.01 sec (ten milliseconds) because pool walls are only so flat; these irregularities in lane length correspond to about five milliseconds at 50m freestyle swimming speeds. Considering the analysis above, a slight current in the pool could easily have as significant an impact on timing round-trip swimming events 100m in length or longer.

CHAPTER 9

Many Rivers to Cross

Introduction

You're resting in slack water along the shore of a whitewater river, wondering how to sneak across the current to get into the next eddy. You're tired after a day of running rapids and wondering, "Gee, what's the fastest way to do that?" And if you're like me, you might ask, "Can I get there in the shortest distance, too?"[49]

Well, sure, you can do that. And whether you realize it or not, you've posed an *optimization problem*. Optimization problems arise all the time in engineering and mathematical physics. They are based first and foremost on a clear, quantitative statement of the desired outcome, such as performing a task in minimum time or with minimum energy, minimizing the amount of metal waste created in forming an automotive part, or transmitting a weak cell phone signal in the presence of interference. After quantifying the goal, the means for finding an optimal way to perform the stated task is usually posed in mathematical terms. Sometimes in very, very mathematical terms involving calculus, matrix algebra, and the like. The river crossing problems posed above lend themselves quite readily to a subset of math called variational calculus.

But who needs that when you can draw pictures instead?

In this chapter, we'll tell you many things you already know, which is always reassuring. And we'll skip the variational calculus. Instead, we'll rely on geometry and trigonometry to

---

49 And I have often wondered, "How do I get across without going for a long, cold swim?"

solve the minimum time and minimum distance crossing problems. We'll employ vectors to model a few cross-stream current profiles problems intuitively.

Vectors and Uniform Current

Let's start with a straightforward example: A river with a current that is uniform across its entire width. As shown in Fig. 9.1, this river has a width of $2M$, which is a convenient way to represent the width. The river flows from bottom to top, with the current vector **C** having magnitude (i.e., speed) C in the vertical (y-axis) direction. The parallel arrows above the x-axis represent the streamlines. Since the current is uniform across the river's width, these arrows have equal length.

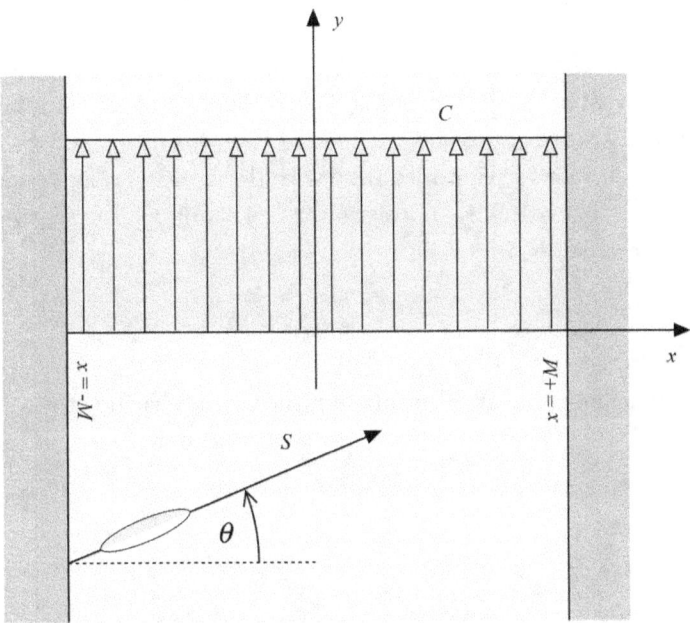

Figure 9.1: River with uniform cross-stream current.

Starting at the left bank, we paddle a canoe across the river at speed S. The paddled hull's keel line is set at an angle θ to the cross-stream (x) direction.

We need to make a careful distinction between *speed* and *velocity*. Speed has magnitude only; it has no direction. For example, we can say that a car is moving at 100 *km/hr* because we read that number on our speedometer. That's a speed, and we refer to quantities that have magnitude but no direction as scalars. Velocity has both magnitude *and* direction; it is an example of a vector. Driving 100 *km/hr* (magnitude) with a N/NE heading (direction) is vector information.

Think of vectors like arrows. An arrow has a length, which is like a magnitude, and points in a direction. We can think of the canoe shown in Fig. 9.1 as moving with a vector velocity **S** shown as the angled arrow. S represents the magnitude of the canoe's velocity (i.e., its speed) while θ represents its direction in the absence of current. Vector notation often uses boldface (e.g., '**S**') to distinguish vector quantities from scalar quantities (e.g., 'S').

We can express vector quantities in terms of their component elements. These components are projections of the vector along coordinate axes, like the x and y axes in Fig. 9.1. The canoe velocity vector **S** may be represented as an ordered pair of its scalar x and y components S_x and S_y as <S_x,S_y>. These are the canoe's velocity elements in each coordinate direction. We order these by convention – the x component followed by the y component – so we know which part is which. The angled brackets are just a notation that lets us know this pair of values is a vector, while the subscripts reinforce which component is which.

In terms of the current-free hull angle θ, using a little trigonometry, the vector **S** is written as

$$\mathbf{S} = \left\langle S\cos\theta, S\sin\theta \right\rangle. \tag{9.1}$$

Similarly, the current is a vector since it has a magnitude C as well as a direction. The current vector may not be particularly interesting since it has only a single non-zero component. However, it still has a direction, and we need to take this into account.

In Fig. 9.1, the current's streamlines align with the vertical (y) axis and are perpendicular to the cross-stream (or "spanwise") direction. As a result, the current vector **C**, expressed in terms of its components, is

$$\mathbf{C} = \left\langle 0, C \right\rangle. \tag{9.2}$$

Equation (9.2) is a compact way of stating that the current has no cross-stream component.

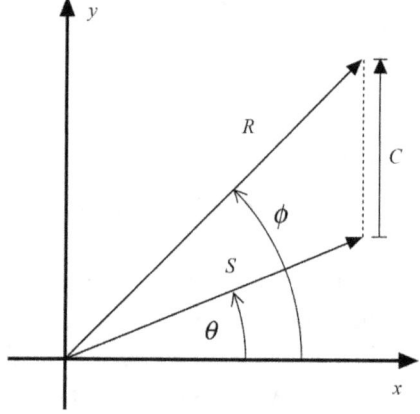

Figure 9.2: Vector addition.

Now recall that the angle θ represents the canoe's heading in the absence of current. How can we determine this angle in the presence of current? We use the properties of vectors and add.

We add vectors pictorially by placing the "base" of one vector at the "tip" of another, making sure to preserve their respective angles as illustrated in Fig. 9.2. The resultant vector **R** is their sum. This resultant vector extends from the base of the first vector to the tip of the second vector, and has magnitude R and direction ϕ. By summing the paddling and current vectors, we see that the current turns the canoe's direction toward the flow. The canoe itself has not rotated! The canoe moves about its center of rotation along the resultant vector $\mathbf{R}$ shown in Fig. 9.2, with an angle ϕ different from the keel angle θ.

We could do a little trigonometry to derive an expression for the resultant vector using the geometric construction shown in Fig. 9.2. But the beauty of representing vectors as ordered scalar pairs is that we can add vectors by merely adding their components. The resultant vector $\mathbf{R}$ is expressed in terms of the canoe and current vectors as

$$\mathbf{R} = \langle S\cos\theta, S\sin\theta \rangle + \langle 0, C \rangle = \langle S\cos\theta, C + S\sin\theta \rangle. \tag{9.3}$$

In vector addition, you merely add the respective scalar components – the x components of the vectors are added and become the resultant vector's x-direction scalar component, and so on. There is no need to break out (more) trigonometry.

We can now determine how long it takes our canoe to cross the river in the presence of this current. We do this by using high school physics and algebra, where we learned that distance traveled is speed times time. The canoe's speed and the river's width are known, so we solve for the crossing time. Since we're only interested in how long it takes to cross the river, we need only consider the x (e.g., cross-stream) component of the resultant velocity vector. That's the direction of motion that gets us across. The cross-stream component of the resultant vector yields our crossing time for the uniform current profile,

$$T_{cross} = \frac{2M}{S\cos\theta}. \tag{9.4}$$

There's something nifty in this result: The time to cross the river *has nothing to do with the current*. It only has to do with the speed that you paddle the canoe in the absence of current and the angle that you set the hull. The angle θ is the angle of the hull's keel line, not the angle of the resultant motion vector ϕ. The river's width $2M$ is fixed, and you maintain a uniform speed S. The keel angle to the far shore is the only variable determining your crossing time.

The hull angle that maximizes $\cos\theta$ will minimize the river crossing time. This is because a larger denominator in equation (9.4) yields a smaller fraction, resulting in a shorter crossing time. Since the cosine function varies in magnitude from 0 to 1, the maximum value you can obtain for the denominator is $S \times 1$. This corresponds to the angle θ equaling 0, which is when the canoe is pointed directly at the far shore. You minimize crossing time for a uniform current profile by putting as much speed in the cross-stream direction as you can.

Now, this probably isn't surprising to many of you; you've paddled before. And you know that paddling this way will get you blown downstream, too! But keep in mind the question we were asking: How to minimize the river crossing time for spanwise uniform current. In any optimization problem, your result is consonant with the stated goal, embodied by the parameter you set out to optimize. If you're seeking minimum crossing time plus something else, that's what engineers call "feature creep." It's a different optimization problem.

A related problem is minimizing the distance traveled for a given paddling speed S and river width $2M$. We can solve the minimum distance problem by seasoning our analysis with a liberal dose of common sense. The shortest distance between where you are on the near shore to the far shore is directly across, a straight line perpendicular to the shoreline. Any other direction entails a longer path length. The only variable is the hull angle θ. So how do we choose the hull angle to minimize distance traveled in the presence of uniform current?

With vector addition, of course.

Recall that the resultant vector **R** is by the sum of current-free canoeing vector **S** and the current vector **C**. It has a component in the cross-current direction and a component in the downstream direction. If the chosen hull angle nulls out the component of **R** in the downstream (y) direction, you won't be blown downstream by the current, and you'll move directly across. Which is another way of saying that you won't travel any distance in a direction that you're moving perpendicular to. Using our ordered pair representation of **R**, this means

$$S \sin \theta = -C . \tag{9.5}$$

In other words, you've set the canoe hull angle so that its downstream component − sinθ times S − cancels the current's magnitude C. The reason they cancel is because of the minus sign. You've chosen an upstream hull angle θ that transforms the vector addition into what's depicted in Fig. 9.3. The current pushes you downstream. With this hull paddling angle, the current won't sweep you away. You're paddling upstream against the current and will travel directly across the river. The resultant travel vector **R** is consequently aligned with the cross-stream direction, and yields in a minimum travel distance. Or as close as you can; the real world is a bit more dynamic, but you get the idea.

Other Current Profiles

A uniform cross-stream current profile may be found in canals and is a reasonable first approximation for some rivers. But what about crossing a river in a bend, where the current often piles up along the outer shoreline? Or a river with variable cross-stream depth, perhaps deepest in the middle but shallow along the banks? Can the analysis above be generalized without using variational calculus?

Well, sure, we can approach these more general current profiles using the tools developed above. And appeal to intuition now that the uniform current profile case is under our belts. We'll begin by unbending a river.

How do you unbend a river? Suppose the river has a bend where the current increases (for example) linearly from the inner shoreline to the outer. In that case, we can imagine taking a cross-stream slice of the river along a radius. Along with that slice, and for some distance upstream and downstream, it looks like a segment of a straight river with the "ramp" current profile shown in Fig. 9.4.

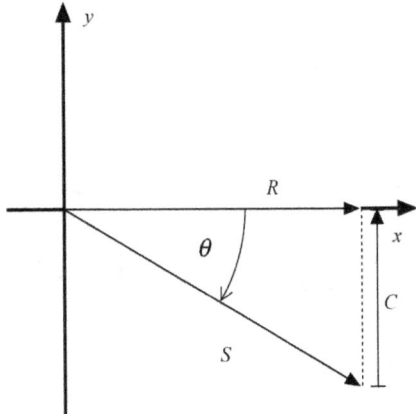

Figure 9.3: Minimum distance crossing resultant in uniform cross-stream current.

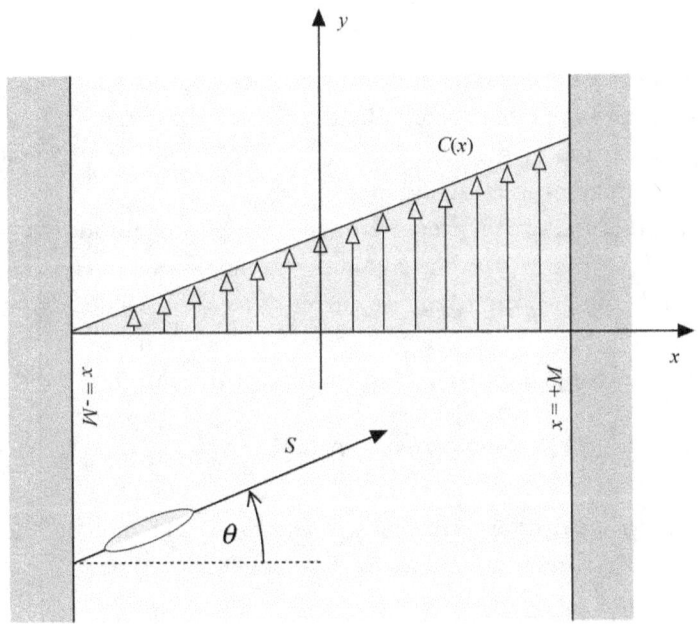

Figure 9.4: River with a linear cross-stream current profile.

If the river is twisty, you'll need to solve the problem in polar coordinates. But for more gradual turns, this is a reasonable approximation.[50] At first blush, this doesn't look much like the uniform cross-stream current case. But if we introduce another approximation (which we'll later dispense with), it does. We'll approximate the current profile of Fig. 9.4 by assuming that over successive spanwise "channels," the current doesn't change all that much. This approximate current profile appears in Fig. 9.5.

---

50 Or at least as good an approximation as assuming a linear spanwise current profile!

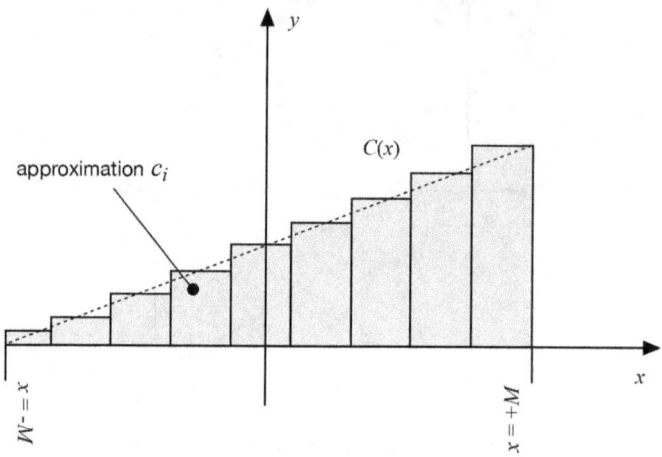

Figure 9.5: Stepwise approximation of the ramp cross-stream current profile.

Here, the current is assumed to be uniform over each adjacent channel. The current magnitude in each is the value of the actual current C at the center of each channel.[51] This stepwise approximation of the function $C(x)$ will be familiar to anyone who has done numerical integration or digital signal processing.

Let's assume that the scalar c_i represents the current magnitude in the i-th step. Consider now the first segment – the one adjoining the side of the river you are paddling from. The first is merely the uniform spanwise current case we considered above, but now for a somewhat narrower "river." We already know how to minimize the crossing distance in that case: choose a hull paddling angle to counteract the current. For this first segment,

$$\mathbf{R}_1 = \langle S\cos\theta, S\sin\theta \rangle + \langle 0, c_1 \rangle = \langle S\cos\theta, c_1 + S\sin\theta \rangle. \tag{9.6}$$

To prevent downstream drift over the first segment, we need to null out the resultant velocity in the streamwise (y) direction. This means over the first segment

$$S\sin\theta = -c_1. \tag{9.7}$$

Using a little trigonometry, you can determine the keel angle θ in that channel via

$$\theta = \arcsin\left(-\frac{c_1}{S}\right). \tag{9.8}$$

For the second step in this approximation, if the current there isn't strong enough to blow

---

51 You can select the current magnitude at either side of each channel or at any location in between as long as you are consistent in doing so.

you downstream, you do the same thing. You already crossed the first segment optimally with no downstream drift and are still heading straight across the river when you start the second step.

As you have probably already deduced, you repeat this process across all steps until you reach the far shore. Nulling the resultant downstream velocity over each i-th step means means for each segment satisfying

$$S \sin \theta = -c_i, \tag{9.9}$$

which leads to a keel angle over each segment of

$$\theta = \arcsin\left(-c_i \Big/ S\right). \tag{9.10}$$

Since the current magnitude c_i in any channel differs from the others, the hull angle in each is different. You are adjusting your hull angle in each segment.

You may have noticed that the analysis in this section hasn't specified the number of steps to approximate the linear cross-stream current variation. You can use as many as you like. Perhaps a lot, a whole lot. The more you use, the better you approximate the actual current profile. As the segments get narrower and narrower, you start to approach a continuum of values. In calculus, this process is called "taking the limit": the channel widths go to zero, or their number goes to infinity. In the limit, at any cross-stream location x, your current-free hull paddling angle must satisfy

$$S \sin \theta = -C(x), \tag{9.11}$$

which leads to an expression for the keel angle at all cross-stream locations x,

$$\theta = \arcsin\left(-C(x) \Big/ S\right). \tag{9.12}$$

The keel angle θ now varies continuously since the argument of the arcsin function varies with cross-stream position. Note that the keel angle depends on the local *ratio* of the current C and the hull speed S, not on their values. Further, the angle does not depend on the river's width.

What does this continuous keel angle variation look like? Pretty much what you'd expect, as seen in Fig. 9.6.

In each plot of Fig. 9.6 the river has been shifted, and scaled to have width $2M = 1$. The cross-stream coordinate extends from the left bank ($x = 0$) to the right bank ($x = 1$); this is just a convenience for computation and plotting. The three cases correspond to ratios of hull speed to a maximum current of 2 (hull speed twice as fast as the current), 1.1 (hull speed 10% faster than the current), and 1.01 (hull speed only 1% faster than the current).

In all three cases, the hull speed is much faster than the current on the left side of the

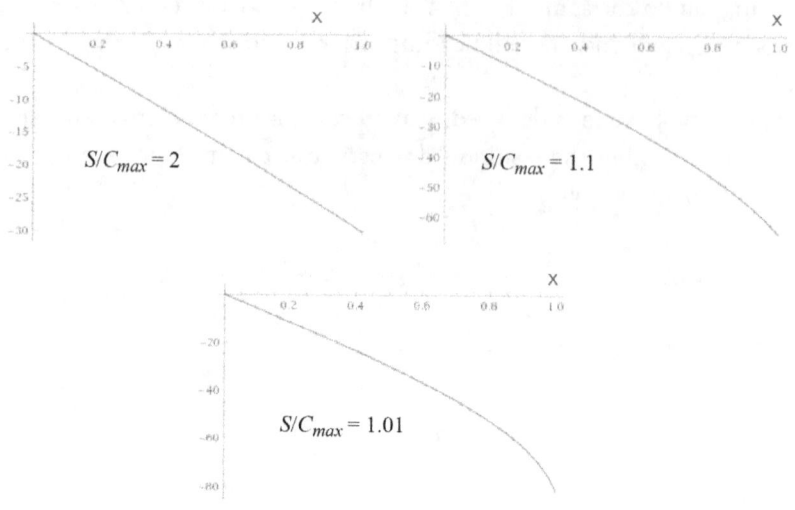

Figure 9.6: Hull angle (degrees) vs. distance for various ratios of hull speed and maximum current speed for the linear current profile.

river. Consequently, the current-free hull angle scarcely varies until you paddle beyond mid-stream. With increasing current magnitude, as you approach the far shore, the canoe must be turned progressively more into the current to continue straight across. Suppose the hull speed is scarcely larger than the maximum current. In that case, the hull is essentially running parallel to shore to resist being swept downstream.

The bottom line: for hull speeds much greater than the maximum current magnitude, for this current profile and others, you can mostly ignore the current variation, set your hull angle, and paddle across with minimum distance. You only need to introduce cross-stream hull angle changes when the maximum current is strong and progressively change that angle as you enter stronger current.

And what about the time to cross for this linear cross-stream current profile? Will the changing current alter our "paddle straight across" approach for minimizing time?

In short, no. Nothing changes. The resultant velocity vector for any cross-stream variation in current magnitude $C(x)$ takes the form

$$\mathbf{R} = \langle S\cos\theta, S\sin\theta \rangle + \langle 0, C(x) \rangle = \langle S\cos\theta, C(x) + S\sin\theta \rangle. \qquad (9.13)$$

Recall that crossing time depends on the cross-stream component of the resultant vector $\mathbf{R}$ for parallel streamlines. For any variable current profile with parallel streamlines aligned with the riverbank, the current does not contribute any "push" across the river, just in the streamwise direction.[52] Consistent with these assumptions, to minimize the river crossing time, always paddle straight across and adjust your hull angle as you go.

---

52 Real rivers have currents that go all over the place. It's just challenging to come up with a representative model for spanwise current that applies to a more than a few special cases.

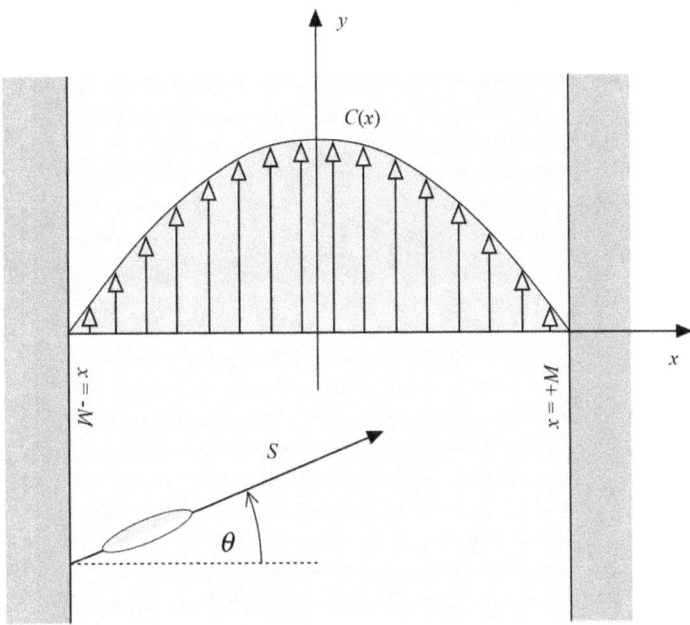

Figure 9.7: River with a parabolic cross-stream current profile.

For the last example we'll consider a river having variable depth. The river's depth is symmetric cross-stream, and the current profile is parabolic. This is depicted in Fig. 9.7.

Once again, we approximate the current profile using a series of discrete "channels" across the river. We assume that each channel has a constant spanwise current, with a magnitude equal to the value of $C(x)$ at its center. The analysis proceeds as before, with increasingly fine slices until we recover the continuous current profile. We already know that the minimum time solution means paddling straight across since the streamlines parallel the river's streamwise axis. Now the variation in keel angle has even symmetry about mid-stream, as shown in Fig 9.8.

In each plot of Fig. 9.8 the river has been shifted, and scaled to have width $M = 2$. The cross-stream coordinate extends from the left bank ($x = -1$) to the right bank ($x = +1$); this is a convenience for computation and plotting. The three cases correspond to ratios of hull speed to a maximum current of 2 (hull speed twice as fast as the current), 1.1 (hull speed 10% faster than the current), and 1.01 (hull speed only 1% faster than the current). The hull speed is much faster in all three cases than the current near the shorelines, so the current-free hull angle is zero as you begin and finish crossing.

With increasing current magnitude near midstream, traveling straight across the river requires the hull to be turned progressively into the current. While the curves in Fig. 9.8 look similar, the maximum angles of the plots' abscissas differ in each case. Near midstream, when the hull speed is scarcely larger than the maximum current magnitude, the hull is nearly running upstream not to be swept away. In each case depicted, the hull angle varies the most when you're furthest from shore. In the field, this makes it harder to choose a keel angle. You're watching both the far shore and the orientation of the keel line. A bit more to manage, but you can do it.

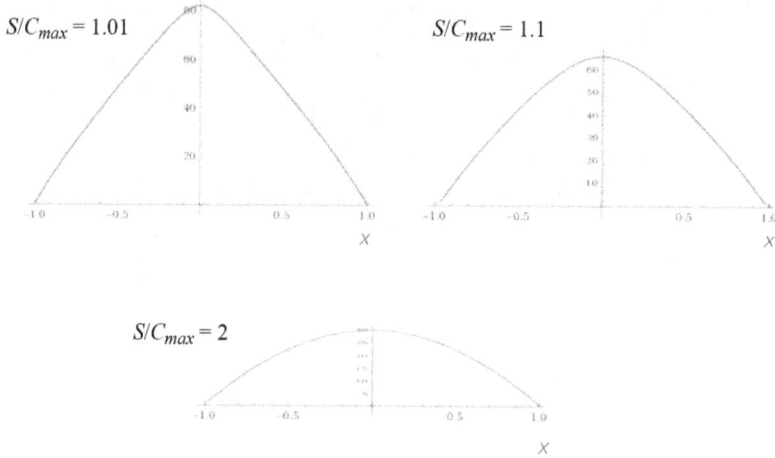

Figure 9.8: Magnitude of hull angle vs. distance for various ratios of hull speed and maximum current speed for a parabolic current profile and minimum travel distance.

You can apply this analysis to the minimum distance problem of starting at one location on the near shore and aiming for a location on the far shore. Just determine the bearing to your target location and use vector addition to solve for the hull paddling angle. This yields a resultant velocity vector **R** coincident with the target bearing. The proof, as they say, is left to the reader.

Take-Aways

- In current, the goals of shortest crossing time and shortest crossing distance are mutually exclusive.
- For rivers with parallel streamlines and no cross-stream current, the shortest crossing time always entails paddling straight across. And for the minimum distance problem, you set the hull angle to counteract the current and travel straight across.
- For more complex parallel current profiles, you can sub-divide the current into several "channels" with uniform current and proceed as before. In the limit, as you increase the number of channels, this discretized approximation converges to the exact solution.

Extra Credit: Two Limiting Cases

What if the current's magnitude C is greater than your paddling speed S? First, the minimum distance solution goes out the window. You can see this by letting the current magnitude be greater than the paddling speed S by an amount dS. Then you can never satisfy the equation above, since

$$S \sin\theta = -C \quad \rightarrow \quad S \sin\theta = -\left(S + dS\right). \tag{9.14}$$

Dividing by S and taking the magnitude of both sides of this equation shows that you would have to find a keel angle θ that satisfies

$$\left|\sin\theta\right| = \left|1 + dS\!\!\Big/\!\!S\right|. \tag{9.15}$$

No angle θ satisfies equation (9.15) for any value of dS greater than 0 because the magnitude of $\sin\theta$ can never be greater than 1. You don't have enough oomph to paddle straight across when the current is that strong!

Another limiting case arises when you decide to call it a day, turn the boat downstream, and head for a takeout on the same side of the river from where you started. This corresponds to a keel angle θ equal to 90 degrees. In this case,

$$T_{cross} = \frac{2M}{S\cos\theta} \rightarrow \frac{2M}{S\cos\left(90\text{degrees}\right)} = \frac{2M}{0} \rightarrow \infty. \tag{9.16}$$

Your river crossing time becomes infinite because you never try to cross the river! Which is obvious in hindsight. But whenever infinities pop up, they are usually trying to tell us something. Either the math's wrong, the problem isn't well-posed, or the infinity needs to be carefully interpreted in light of reality. As to whether it takes an infinite amount of time to cross a river you don't paddle, who's got the free time to check?

CHAPTER 10

Trim

Introduction

Have you ever wondered why you sometimes struggle to control your canoe when underway? Have you asked yourself why you should distribute weight in a hull to affect a particular trim? Have you felt like your speed was lacking in certain water conditions? This chapter uses physics to understand the impact of changing your hull's underwater shape via trim. Correct trim can facilitate improved control, speed, turning, and efficiency when paddling.

Trim

Trim is the fore-to-aft inclination or declination angle of a hull along its length. If a hull has no rocker,[53] a spirit level placed on its keel line indicates the trim angle. For a hull with a symmetric fore-to-aft rocker, you place the spirit level on the keel line's tangent at the center of the hull.[54] For the sake of simplicity, we'll call the angle the "trim angle," or simply "trim," and we'll call a horizontal trim angle "level."

There are two types of trim: static trim and dynamic trim. Static trim is the angle when the boat is in still water and not moving. Dynamic trim is the angle when the boat

---

53 Rocker is the front-to-back curvature of a hull's keel line. If the keel line is straight (neglecting any curves where the keel and stems meet) the hull has no rocker. Rocker is usually symmetric about the hull's center. Specialized hulls may have more rocker in the front or back half.

54 For a hull with an asymmetric rocker, look for a waterline, but you may have to call the hull's manufacturer!

is underway at a steady speed. Dynamic trim can change as speed increases up to the hull speed. These two types of trim are illustrated in Fig. 10.1.

In Fig. 10.1(a), the stationary hull's trim is level. When underway at its hull speed, a bow wave and sternward trough form alongside the hull. As we learned in Chapter 3, the bow slows the water as it encounters the hull. Adopting a paddler-centric coordinate system, the kinetic (motion) energy of the inflow must therefore decrease. Since energy is conserved, the forward flow's energy transforms from kinetic to potential energy in the form of a bow wave. And because the water accelerates as it passes alongside the hull, the kinetic energy downstream of the bow increases. Consequently, the water's potential energy decreases toward the stern. The drop in potential energy results in the sternward trough alongside the hull shown in Fig. 10.1(b). This combination of a bow wave and a sternward trough causes the bow to rise. The hull is buoyant – remember Archimedes' Principle? – and inclines itself to match the inclined dynamic waterline you can draw from the bow wave to the sternward trough. Consequently, the hull in Fig. 10.1(b) has a "bow up" dynamic trim.

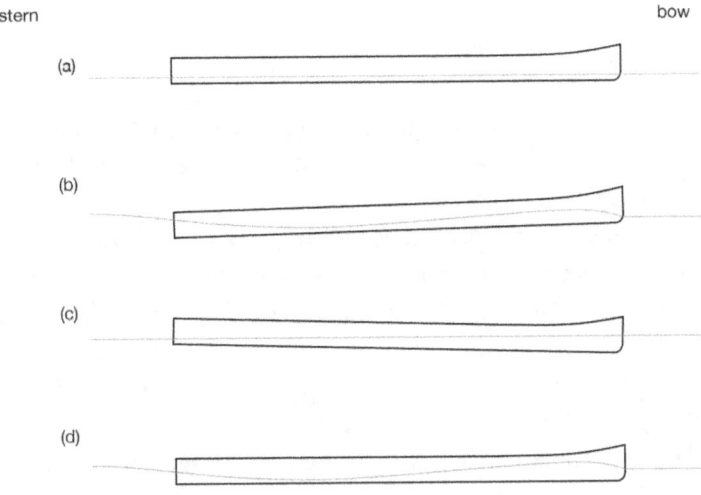

Figure 10.1: Static (a), (c) trims, and corresponding dynamic (b), (d) trims.

In Fig. 10.1(c), the stationary hull is trimmed bow down. When underway, the bow still rises, and the stern still sinks because the flow past the hull slows at the bow and accelerates toward the stern. However, since the hull has a bow down static trim angle, once underway the bow down static trim becomes a level dynamic trim, as shown in Fig. 10.1(d). We've picked a static trim angle that exactly offsets the change in trim once underway. Since hulls are most efficient with a neutral dynamic trim, it behooves you to experiment with fore-to-aft weight distribution to find this trim point or one close to it, and then vary your trim depending on conditions. For example, very shallow water entails a larger bow wave, requiring a more aggressive bow-down static trim to keep the hull level and not "climbing a hill" in shallow water waves.

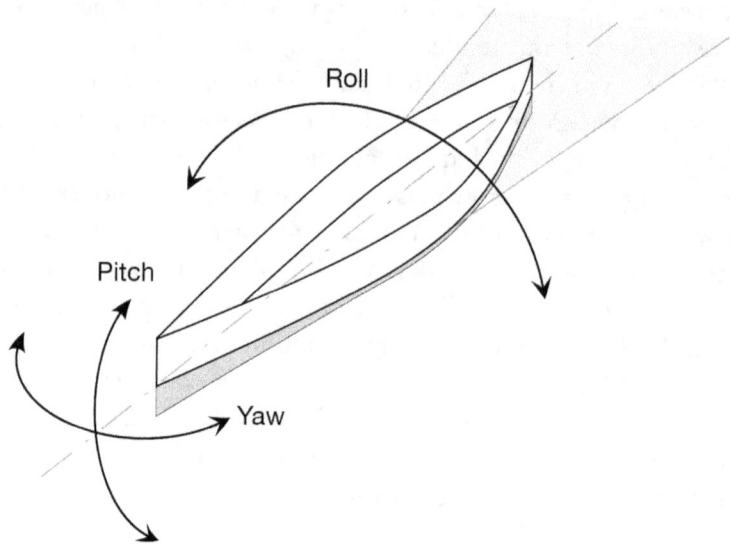

Figure 10.2: Roll, pitch, and yaw.

Trim also impacts a hull's yaw handling, e.g., side-to-side motion, as defined in Fig. 10.2. Consider the three trim scenarios depicted in Fig. 10.3: level, bow-down, and bow-light (or stern-heavy if you prefer).

Experientially, many, if not most, readers will know that a bow down trim "lightens" the stern and makes it easier to follow a line on a twisty river when running upstream. Or that a bow light trim helps in current when back-paddling to "set" a hull to a particular cross-river location. If you trim the bow down and encounter a beam wind, the hull will turn into the wind, while a bow light trim causes the hull to turn away from the wind. We all know this to be true. But why does it happen?

We'll start with a simple experiment that you can do right now as you're reading this chapter. From Chapter 1:

> Straighten your arm out in front of you, with your hand flat and parallel to the floor.
>
> Now briskly move your hand side-to-side, arm straight, and notice the sensations of air passing it.
>
> Next, repeat this experiment with your hand still flat but now held vertically (i.e., perpendicular to the floor), briskly moving your hand side-to-side, arm straight. Notice anything different? Do you feel more force distributed over your palm in this orientation than before?

This simple experiment is even more dramatic if you stick your hand in the water alongside a moving hull. With your hand aligned to the flow, water rushes past it. With your hand held perpendicular to the flow, the amount of force on your hand can be surprisingly large. You can generalize this experience from hands to hulls. Why does it take less paddle force to move a hull forward compared to drawing it sideways? If you said friction, note that the friction drag force is proportional to the hull's wetted surface area – see the Drag Equation in Chapter 1. The wetted area doesn't change whether you paddle a hull forward, draw it sideways, or do anything in between. While friction drag is present, there's something else going on: form drag.

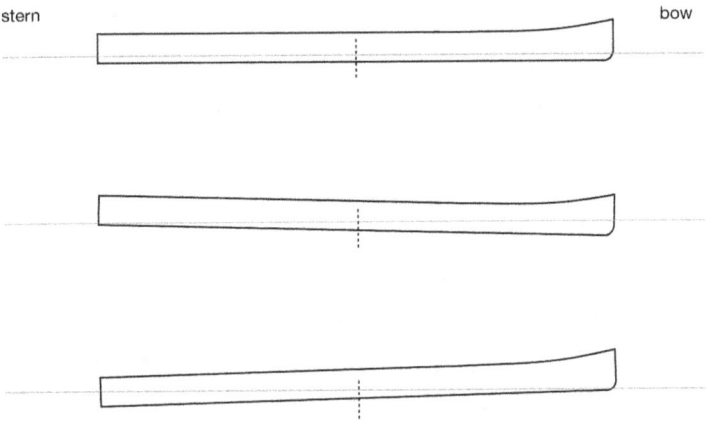

Figure 10.3: Three states of trim for a hull without rocker.

Form drag arises because objects in flow generate wakes, which causes the flow's momentum to change. Changes in momentum give rise to forces due to Newton's 2nd Law. And as we saw in Chapter 2, that drag force is proportional to the wake's cross-section area, and the wake's area is proportional to the object's cross-section area.

Returning now to Fig. 10.3, note the vertical dotted line amidships in each trim scenario. The underwater hull cross-sectional area viewed from the side is pretty much equal for the hull's front half and its back half for level trim. For bow-down trim, the front half's submerged cross-section is greater than the back half. Consequently, the hull's submerged front half has greater form drag in yaw than the back half. Fundamentally, drag is resistance to movement. So a bow-down trim implies that the bow will resist side-to-side movement more than the stern because of its greater form drag. For bow-light trim, the submerged back half of the hull has greater form drag in yaw than the front half. The stern will resist side-to-side movement more than the bow because of its greater form drag.

A rockered hull's side-on form drag is similar, but the effect may not be as dramatic as a straight-keeled hull. You can see why in Fig. 10.4 for a symmetric rockered hull. A rockered hull has less volume in the water toward the stems, and more in the belly. When you trim a rockered hull bow down, the difference in side-on submerged cross-section area, front half to back half, is less than for a straight keel. This difference becomes less with increasing rocker.[55]

You might conclude that a hull's deeper end acts as a rudder. This isn't the case. A rudder rotates to deflect incoming flow that generates a side force from the dynamic pressure,

---

55 The effect is still present in a rockered hull, just slightly less than for a straight-keeled hull. If the submerged rocker's shape were perfectly circular (and the hull had a uniform cross-section bow to stern), the effect would disappear – but that would make for a fairly strange hull.

and lift. This pulls a rudder-equipped vessel's stern to one side or the other. Unless our hulls bend,[56] its lower end doesn't rotate to deflect flow. Instead, the submerged portion acts more like a pivot. It does not act like a fixed pivot, e.g., a hinge. But the hull's deeper portion resists side-to-side motion compared to its shallower end.[57]

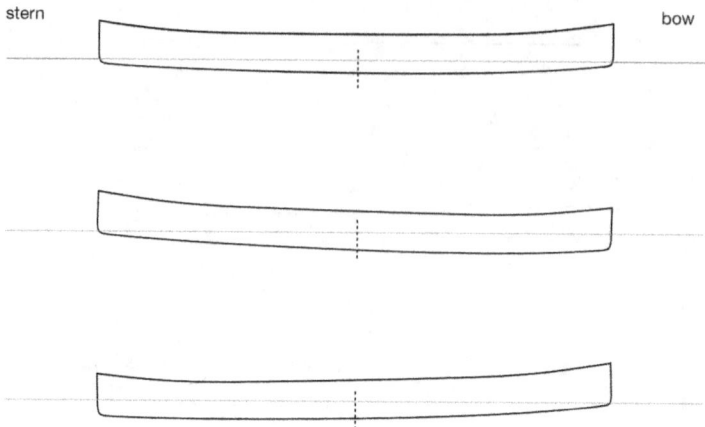

Figure 10.4: Three states of trim for a hull with rocker.

If the hull is underway with a bow-down dynamic trim, the pivot point keeps the hull aligned with the flow. If instead, the hull has a bow-light dynamic trim, the bow may feel soft. The boat may track poorly and even spin in the current. A bow-light dynamic trim

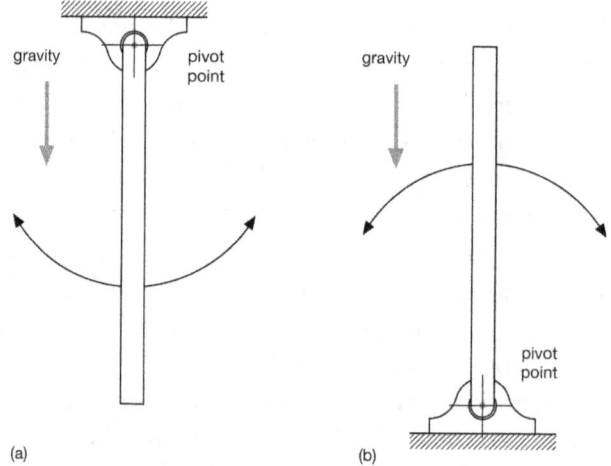

Figure 10.5: Stable and unstable pivot configurations analogous to bow-down (a) and bow-light (b) trim.

---

56 Which sometimes happens, but usually when it wraps around a fixed obstacle and breaks.

57 And the fixed end of a hinge's pivot resists motion.

is a bit like a pendulum in configuration 'b' illustrated in Fig. 10.5.[58]

When a hull is stationary and has neutral trim its pivot point is coincident with its center of mass. When a hull is underway, the incoming flow pushes against the hull. The dynamic pressure force produced by the incoming flow is balanced by the propulsive force moving the hull forward. These two forces must strike a balance to maintain stability, thus the pivot point moves forward of the center of gravity. This directional stability is further enhanced by a bow-down trim. A dynamic bow down trim is analogous to a pendulum hanging in a gravity field, as depicted in Fig. 10.5(a); the forward portion of the hull acts as a pivot. The acceleration of gravity induces a force that pulls downward on the pendulum, acting as a restoring force to align it with the gravity field when deflected to either side. The hanging pendulum is stable because the pivot point is at the top: the force induced by gravity acts below the pivot through the pendulum's center of mass. It "wants" to align itself with gravity, much like a hull with bow-down trim wants to align with incoming streamlines.

By contrast, Fig. 10.5(b) depicts the same pendulum in a gravity field, with the pivot point at the bottom. This configuration is analogous to a hull underway into incoming flow with a bow-light trim. Gravity induces a force that pulls on the pendulum through its center of mass, which is now above the pivot. The pendulum in Fig. 10.5(b) is only aligned with the gravity field when it is perfectly balanced on the pivot. Consequently, this configuration is unstable and prone to rotate around the pivot. Sound familiar?

You're familiar with this inverted pendulum problem If you've ever tried to balance a pencil on your fingertip. Only through continuous adjustment can you balance the pencil.[59] Similarly, a hull that is bow light with its pivot more astern, when moving into an incoming flow, can be destabilized and knocked off a straight course unless you introduce corrective paddling strokes. In strong current, this effect is rather dramatic and remarkably quick. But for a skilled whitewater paddler, this slight trim-induced instability can be employed for quick maneuvering.[60] Or, wilderness paddlers working a loaded boat down rapids will use a bow-light trim while back-ferrying (aka, setting) to move across a current and avoid downstream obstacles.

Lean

In addition to trim, a hull's lean impacts handling. Skilled SUP paddlers move all over their boards to carve turns and spin their craft. They are both inducing lean and significantly changing the combined board/paddler system's mass distribution to create pivot points and new moments of inertia. In what follows, we'll restrict ourselves to some basic hydrodynamics of leaned hulls. But keep in mind that, in addition to trim, a hull's mass distribution affects how it turns.

---

58 Note that a hull's pivot point is not fixed like the bearing shown in Fig. 4; this is an analogy. This mechanical analogy becomes more realistic if we let the pivot translate side-to-side or fore-and-aft.

59 The inverted pendulum models the canonical feedback control problem central to rocket flight.

60 Another analogy is that modern high-performance jet aircraft are inherently unstable and are stabilized via their flight control feedback systems. Yet this instability is used to make exceedingly quick climbs and turns, faster than a more stable airframe is capable of, or to produce decoupled maneuvers such as horizontal flight while maintaining downward trim.

Consider an un-rockered hull in a uniform flow with streamlines, as shown in Fig. 10.6.[61] Because the hull is symmetric side-to-side, the streamlines are symmetric across the hull as well.

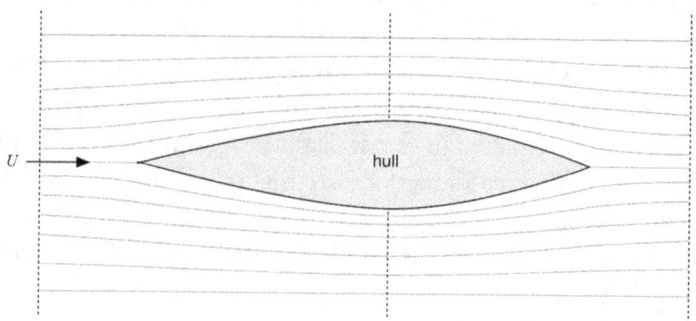

Figure 10.6: Hull and inviscid streamlines for a given vertical station.

In Chapter 3, we learned that the incoming flow's kinetic (motion) energy increases at the hull's maximum width. The vertical dotted line indicates this location in Fig. 10.6. Increased flow speed means that the water's kinetic energy is larger than the inflow's kinetic energy. Because total energy is conserved, this increased kinetic energy requires that the potential energy at this location be less than that at the bow or stern. For fluids, potential energy means a change in pressure. Thus, the flow's pressure at the hull's maximum width is less than at the bow or stern; the pressure varies along the hull's length along with the changes in flow speed at each location.

Since the hull is symmetric side-to-side, the pressure changes are also symmetric side-to-side. Consequently, the pressure-induced forces induced along the sides of the hull balance each other out. With no net cross-force, the hull doesn't accelerate abeam, in accord with Newton's 2nd Law of motion. Recall that objects accelerate in the presence of unbalanced forces.

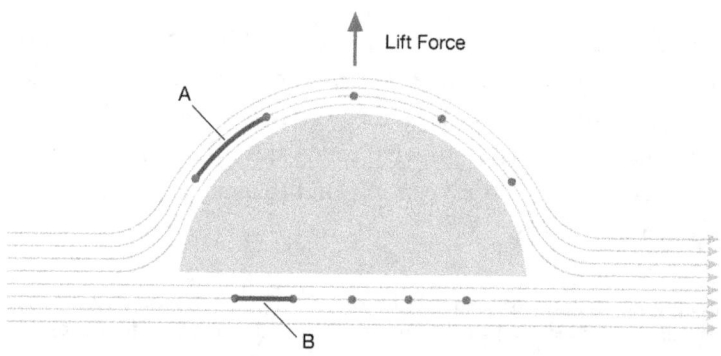

Figure 10.7: Hemispherical section and inviscid streamlines.

---

61 This is a Galilean transformation from an inertial reference frame to a paddler-centric reference frame.

What if the hull in Fig. 10.6 is asymmetric below the waterline? This is analogous to flow past a hemispherical rod, as shown in Fig. 10.7. An object with a hemispherical cross-section is asymmetrical to this inflow. The streamlines in Fig. 10.7 move from left to right. Note the dots are shown in one streamline below the rod and one streamline above. Initially, these dots were parallel to each other and uniformly spaced in the streamwise direction before encountering the rod. As the flow moves past the lower half of the rod, these streamline markers retain their uniform spacing; the flow there does not accelerate. For inviscid flow, these streamlines rejoin downstream of the rod. But the flow over the top of the cylinder entails a longer path than below, as indicated by the path segment lengths A and B. Consequently, the marker dots spread out over the curved top half of the rod. By covering a greater distance in the same amount of time as the flow below the rod, the upper flow is faster and thus has greater kinetic energy than the flow below.

Because energy is conserved, this increased kinetic energy necessitates a decrease in potential energy there. Consequently, the pressure along the curved half of the cylinder is less than along the flat half. This force imbalance gives rise to a net lift force on the rod, as indicated in the figure. Since unbalanced forces accelerate objects due to Newton's 2nd Law, the cylinder will move in the lift force's direction.

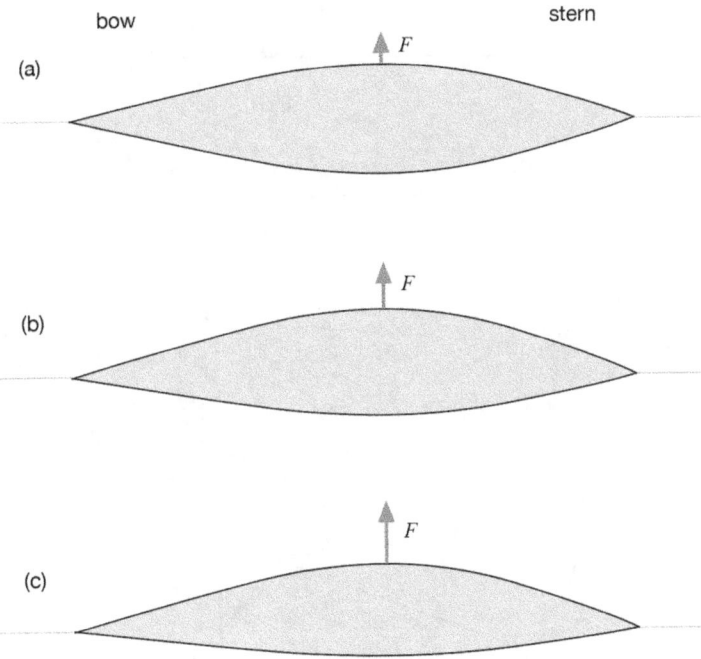

Figure 10.8: Waterline shape for various leans, plan view, non-rockered hull. The horizontal lines connect the points where the stems and water surface meet and are not streamlines.

When we lean a hull to one side, its shape below the waterline becomes asymmetric. This is shown in Fig. 10.8 for Fig. 10.6's hull and three leaned states. As the hull is progressively leaned, we expect the net sideways force F to increase as well due to the increasing asymmetry, as suggested in the Figure.[62]

If the hull has a pivot owing to trim and/or motion, this unbalanced force causes the hull to rotate about that pivot. This is why we lean a hull to the outside to carve a bow-down turn; the hull will skid toward the leaned side and rotate owing to the sideways force and the lightened stern. Add in a brisk forward stroke or a sweep on the leaned side, and you're turning.

The crosswise unbalanced force of a leaned hull does not itself turn the hull. If the pivot location is offset from the side force's point of action, this creates a torque. This torque results in rotation, as noted in the example above. If the pivot and force are coincident the hull sideslips.

A leaned hull also presents an asymmetric head-on profile to the flow, as shown in Fig. 10.9 for various amounts of lean. There is an outward-directed unbalanced force due to the asymmetry below the waterline. And the hull's front-on form drag becomes asymmetric as well, just like we saw with front-to-back trim. Greater form drag on the deeper side creates an "outside pivot" that encourages the leaned hull to spin. An outside lean and a bow-down trim reinforce each other's induced form drag to turn a hull.

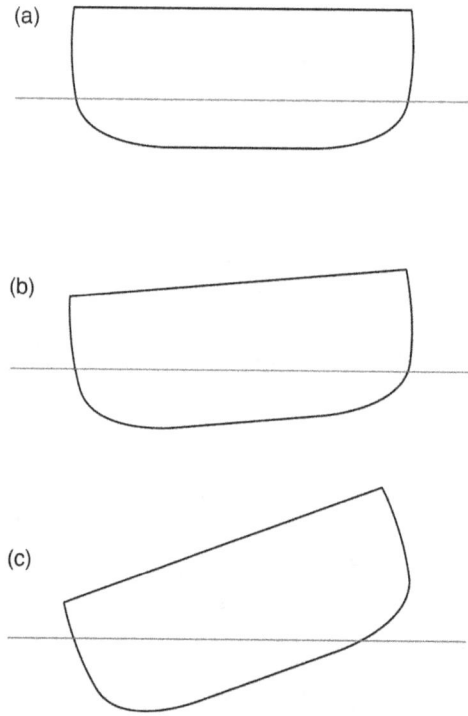

Figure 10.9: Three hull leans, front view.

---

62 Note that for simplicity we are neglecting free-surface effects that arise from the lean.

Turns

We can now analyze the forces acting on a hull that enters a turn starting from full stop. This is depicted in Fig. 10.10.

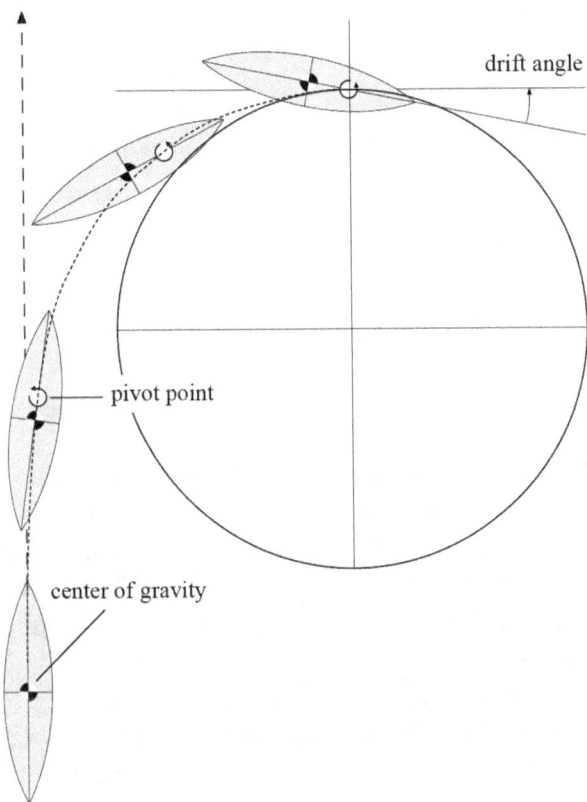

Figure 10.10: Hull entering turn from a full stop.

When our hull is stationary its center of rotation and center of gravity are coincident. As we paddle the hull up to speed its pivot point transfers forward. The paddler(s) initiates the turn through a combination of outside- and bow-down trim and turning strokes. A tandem canoeing team may elect to have both paddlers paddle on the outside to encourage turning while maintaining speed. As we've all experienced, the hull moves through a turn in a kind of skid or drift. When the desired turning radius is established the keel line is deflected inward from any tangent to the turning circle. This angular offset is the *drift angle*.

In the absence of wind or current, and with a constant bottom depth, if we maintain trim and propulsion the hull carves a circle. This is because the various forces acting on the hull are in balance. This is illustrated in Figure 10.11.

First, the forces are depicted as vectors acting through various locations on the hull. For the sake of visual clarity, the lengths of these vectors do not quantitatively reflect the actual magnitudes of the respective forces.

Because of the drift angle the bow is constantly being deflected by an inflow of water.

This inflow "piles up" on the outside of the bow. The resulting dynamic pressure creates a force directed inward at the bow. The distance from the center of this dynamic pressure to the pivot point is a moment arm; the dynamic pressure force is trying to turn the hull inward around the pivot point.

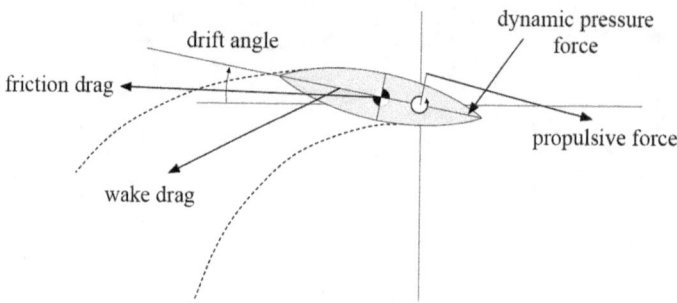

Figure 10.11: Forces acting on the hull in a turn (not to scale).

We idealize the paddler(s) propulsive force as acting (mostly) parallel to the keel line. This too acts via a moment arm from the pivot point; the moment arm is physically comprised of the paddler(s) torso(s), arms, the seat(s) and mount(s), and hull structure.

Friction drag retards the hull's forward movement. Since friction drag is related to the hull's surface area, and the surface area stays the same despite the hull's drift, we show it acting through the center of mass.

Finally, the wake of a turning hull moves along with it. However, the wake is skewed toward the inside of the turn. As we learned in Chapter 2, the wake's velocity deficit induces a drag force on the hull. Since the wake is skewed inward, this drag force no longer acts along the keel line. Instead, wake drag acts from a point behind the hull's center of gravity, directed toward the wake's center. This force acts over a moment arm from the pivot point.

When considered in total these forces, acting through their moment arms, strike a balance around the pivot point. If the trim and propulsive force are held constant the hull will turn at a constant radius until the cows come home. Changes in any of these forces will disturb this equilibrium until a new balance is achieved. Finally, by eliminating the outward lean and transitioning to forward strokes, the hull will exit the turn.

Take-Aways

- Now you know how to trim your hull for efficient paddling in a variety of flow and wind conditions.
- Trim induces asymmetric form drag and inviscid pressure forces due to hull asymmetries below the waterline.
- Trimming a portion of the hull downwards creates a pivot point around which the hull can turn.

- Combining a bow-down trim with an outside lean creates both a forward pivot and outward-directed forces to help a hull carve turns.
- Constant-radius turns embody a balance of propulsive, dynamic pressure, and drag forces.

Further Reading

Laurie Gullion, *Canoeing and Kayaking Instruction Manual*, American Canoe Association (1987).

Bill Mason, *Path of the Paddle*, NorthWord Press, Inc. (1984).

CHAPTER 11

A Real Drag

Introduction

I've often wondered if we can characterize the performance of a canoe, kayak, or SUP using just a few numbers, like the information printed on a new car's window sticker. We see ads for various hulls touting their performance in waves, wash, shallow water, deep water, following seas, etc. You read accolades like fast, stable, and "handles X with ease" for whatever 'X' is. But this begs the questions: fast, stable, and with ease compared to what? Wouldn't it be helpful if we quantified fast on a scale from 1 to 100?

Now granted, this perspective is itself limiting. There are qualitative, paddler-specific characteristics that factor into hull selection and use. I've climbed into canoes that I can't keep upright, while other paddlers thrive in them. The fact that certain hulls are fast is irrelevant if I can't handle the boat! But still, I'd like a number, please, if only to convince myself that a boat I'm buying is a step up from what I already own.

It's not surprising that physics provides one helpful number for characterizing and comparing hulls. And equally as important, we can derive this number experimentally to characterize our craft. This parameter quantifies what holds us back from paddling faster or with ease: the drag coefficient.

Modeling a Hull in Motion

We model a hull in motion using the balance of forces acting upon it. The paddler exerts a propulsive force F_p, which via Newton's 2nd Law is proportional to the combined mass m

of the paddler(s), hull, and gear times their acceleration. The propulsive force is opposed by a drag force, as depicted in Fig. 11.1. Combining these three effects leads to our governing equation of motion

$$F_p - C_D v^2 = m\frac{dv}{dt}.$$ (11.1)

Here, v is the hull speed, and C_D is the hull's dimensional drag coefficient. The derivative dv/dt is the rate that speed changes, otherwise known as acceleration. The drag force opposes the propulsive force, hence the minus sign. As we learned in Chapter 1, the drag force is proportional to the square of hull speed. Note that this model omits the effects of wind and current. We can include these, but for simplicity will leave them out. Since our analysis is limited to the horizontal direction, all quantities are scalars. The propulsive force and the speed also vary with time t. The other terms in the equation are constants.

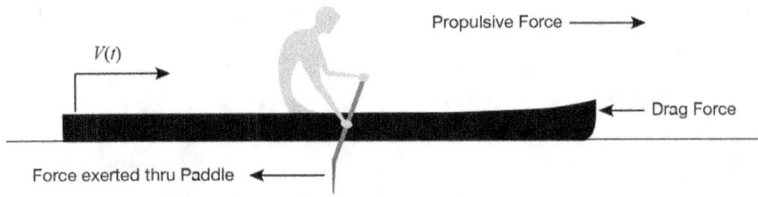

Figure 11.1: Idealized paddler and dynamics elements.[63]

As noted in Chapter 1, the drag force on a canoe, kayak, or SUP can be characterized – subject to a few assumptions – by this single parameter in Equation (1.15). The drag coefficient is a property of the hull and how it sits in the water.[64]

The governing equation above is a Ricatti equation which we can solve in closed form[65] for certain paddling force functions F_p. If a hull is moving with a known speed V_0, and at time $t = 0$, we stop paddling,[66] it's straightforward to solve this equation. The "homogeneous" (e.g., zero propulsive force input, or "unforced") solution for the speed v as we coast is derived in the Extra Credit section. It takes the simple form

---

63 Note that the propulsive force is a reaction force, acting in the opposite direction to the force exerted through the paddle. This is due to Newton's 3rd Law, which requires a reaction to every action. Also, the propulsive force must point in the direction of motion. Otherwise, the propulsive force would not accelerate the hull in the forward direction.

64 When you climb aboard a canoe, kayak, or SUP, your weight causes the hull to displace water, pushing the keel further below the water's surface than when unloaded. In practice, the drag force reflects how the hull sits in the water, e.g., on the actual shape and volume of the hull in the water when paddled.

65 We can write down a closed-form solution on a piece of paper without resorting to numerical simulation.

66 One of the great things about classical physics is that you can "rewind the clock" and make the starting time equal to anything you want. Setting it to zero is mathematically convenient.

$$v(t) = \frac{V_0}{1 + \left(V_0 \dfrac{C_D}{m}\right)t} \cdot \qquad (11.2)$$

You can verify that this solution is correct by substituting it into Equation (11.1) with F_p = 0. Qualitatively, the expression for speed $v(t)$ in Equation (11.2) makes sense. When time $t = 0$, the hull speed equals the initial speed V_0; check. As time t increases, Equation (11.2)'s denominator gets larger, indicating the hull speed v decreases over time. Check! This slowing down reflects our everyday experience of our craft.

So, where do we find values for the drag coefficient? You most likely can't look it up on your favorite manufacturer's website. However, we can use Equation (11.2) to measure it ourselves. The homogeneous solution requires the initial speed V_0. We can measure the speed over time using a GPS receiver. The homogeneous solution also requires the combined mass of the hull, paddler(s), and gear, which we'll measure using a bathroom scale (or two). Then, we can vary the Drag Coefficient C_D in Equation (11.2) to produce predicted speed versus time curves. We deduce the drag coefficient C_D when our predicted and measured speed versus time curves match.

Let's try it and find out.

Taking It To The Water

The model above suggests the tests we'll conduct. The solution to the homogeneous governing equation comprises measurable quantities: the initial speed V_0 and the combined mass m. It describes a scenario where you first paddle to bring the hull up to speed, then stop paddling and coast. If you measure the speed v as a function of time, you can plot this data, compare it to the model's prediction, and use that comparison to deduce the drag coefficient. Your hull will then be characterized.

To get test data, I first had to select a hull. I chose my Savage River D-IIx USCA C-1 solo canoe. You can use any hull you want if you follow the test protocol outlined below, including SUPs, tandems, C-4s, etc. During all tests, I set the seat's fore-and-aft position to provide a neutral static trim.[67]

I also required an appropriate venue. Our goal is to collect data that reflects the intrinsic hydrodynamic characteristics of the hull, not shallow water effects. Consequently, the water at the test site had to be as deep as half the hull's length; this is the deep water transition depth derived in Chapter 3. Since my C-1 is 18' 6" (5.64m) long, I needed a course that was at least 9'3" (2.82m) deep. To get the hull up to speed and then coast for a sufficient time, I estimated the course needed to be about a tenth of a mile (160m) long. The no-current requirement indicated that a lake or pond with appropriate depth was the right choice. And finally, the requirement of no wind meant I conducted tests very early in the morning. The test site, shown in Fig. 11.2, met these requirements.

After searching fishing depth maps for nearby lakes and ponds, I found a local pond that met my criteria. I've often used it for interval and other short-duration workouts.

---

67 You can use the test protocol described here to study the influence of trim for a given hull parametrically.

Figure 11.2: Test site at 5:30 am.

You'll find the pond's depth map in Fig. 11.3 – the test course is highlighted on the map. The depth ranging from 8' to 11' over the course met my depth requirement, but barely. The test area had an open view of GPS satellites. Test durations were short enough that a single

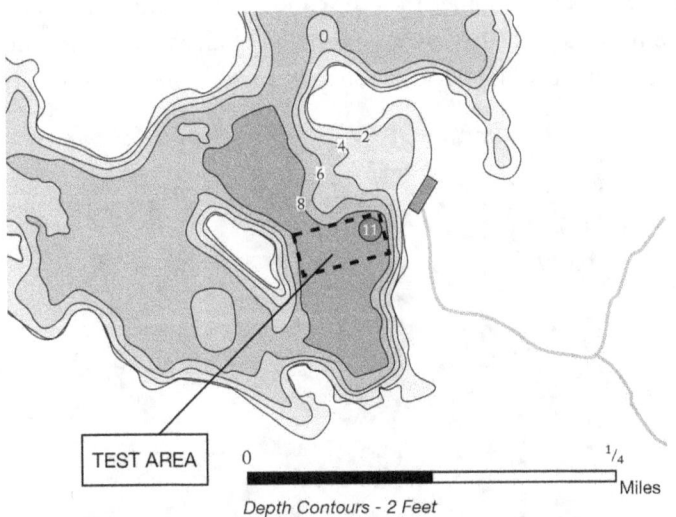

Figure 11.3: Pond depth map and test area boundaries.
(Data source: Massachusetts Division of Fisheries and Wildlife.)

GPS satellite constellation could constantly be in view. There wouldn't be any "jumps" in position due to a changing satellite constellation.

Paddle in hand, I pushed off from the shore, fired up my GPS receiver, and waited for it to lock onto the satellites. I warmed up a bit then paddled to the test area. The pond was flat calm, and aside from a fisherman I saw on shore I had the venue to myself. Recent rains had topped up the pond's level nicely. Tests were conducted as follows:

- After several practice runs, I performed a series of eight (8) test runs. Each test lasted approximately one minute and ten seconds (1:10) and covered about one-tenth of a mile.

- Test runs alternated in direction to overcome any systematic bias that testing in one direction alone might introduce.

- Data was acquired using a Garmin Forerunner 920xt. I configured this sports computer to acquire one datum set (time, position, speed) per second. Note that many wrist-mounted sports computers employ "smart" data recording. Smart data recording entails taking fewer data points, further apart in time, and at irregular time increments.

- Each test started from a stationary position. I started the GPS receiver's data recording, then brought the hull up to speed over 15 – 20 seconds.

- Once up to speed, I stopped paddling, sat upright as if I were paddling, and feathered the paddle blade. Feathering minimized drag conributions from blade aerodynamics.

- Once the receiver displayed zero (0) speed, I stopped the GPS receiver's recording and saved the data.

I discovered that it's tough to keep the hull perfectly vertical when coasting at low speed, with no roll or wobble whatsoever. Sometimes the canoe started carving a turn when I stopped paddling. Or the hull started turning approximately halfway through the coasting phase, as shown in Fig. 11.4. In Fig. 11.4, the canoe was traveling right-to-left; the white trace

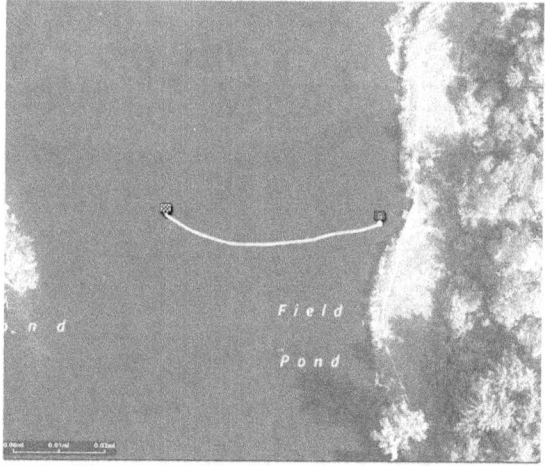

Figure 11.4: GPS track, Test 1. (Map data ©2019 Google)

reflects the GPS track. This carving occurred toward either side; there was no consistent direction attributable to hidden currents or a brief puff of wind.

You'll find a representative trace of recorded speed vs. time in Fig. 11.5. Since the GPS receiver acquired and stored data once per second, we don't see any of the small speed variations associated with each stroke before the coasting phase. This would require a different GPS receiver and software capable of acquiring and recording data 5-10 times per second, which I did not have. Fortunately, the coasting phase speed trace is a generally smooth curve, as seen in Fig. 11.5, so the once-per-second data record was sufficient to capture the dynamics. Also, this receivers' speed data below ~1 mph is suspect; note the immediate jump in the data at the start. Above 1 mph GPS data for this receiver is accurate to about 1-2 tenths of a mile per hour at best. Fortunately, I had a good lock on the satellite signals, and the data looked reasonable given prior experience with GPS.

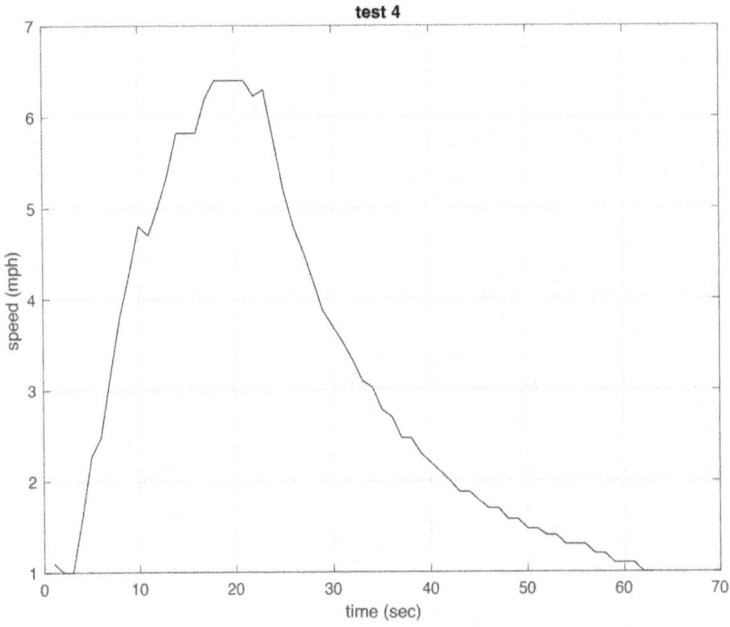

Figure 11.5: Representative speed vs. time data plot (test 4).

Data Reduction and Analysis

After acquiring data from eight runs, I crunched the numbers. In reviewing the test results, I found that it was difficult to discern when my paddling stopped, and the coasting phase began. In part, this is attributable to the GPS data points' accuracy, as noted above. Further, the data was recorded once per second. The "stop paddling" moment could coincide with one data point, the following data point, or some time in between. The solution to the homogenous governing equation requires the initial coasting speed V_0. However, the paddle force is still zero for all times after I stopped paddling – the governing equation is unforced (e.g., homogeneous). If I select any "first" data point after the hull starts coasting, the homogeneous solution applies. I need not select the exact moment paddling stops.

To derive the drag coefficient, I selected data from two consecutive runs in opposite directions. These tests had minimal turning compared to the other test sets. GPS tracks from these tests appear in Fig. 11.6.

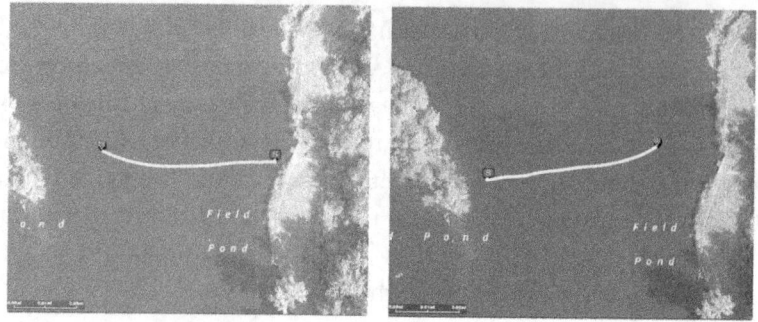

Figure 11.6: GPS tracks, tests 5 and 6. (Map data ©2019, Google).

Each data set comprises a series of data at time points t_i, where 'i' is an index that denotes a specific data point with corresponding speeds v_i. The model solution predicted speed data at these time points given the total mass m, the initial speed V_0, and an estimated value of the drag coefficient C_D. Values of the drag coefficient were varied in the model to minimize the normalized mean square error e between the test data and the model prediction; see the Extra Credit section for details. The comparison between predicted speed values and data from tests 5 and 6 are presented in Fig. 11.7.

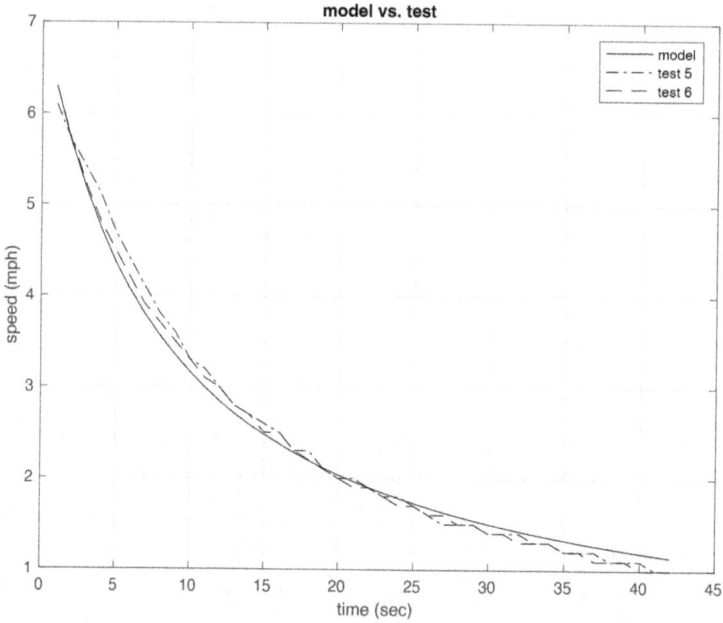

Figure 11.7: Speed vs. Time during coasting phase, model vs. test data.

As seen in the figure, the agreement between theory and experiment is quite good. The difference is attributable in part to assumptions inherent in the model. We have assumed that the total drag force is proportional to the square of the speed alone, lumping all drag effects into a single parameter. Alternate approaches add drag model elements linearly proportional to speed, or proportional to a fractional power of speed. These additions might improve the model fidelity to match the data better but require a data regression to infer multiple drag parameters. They also entail a more complex dynamic model. Our simpler model works quite well and offers the advantage of a closed-form solution to the governing equation.

The mean square error was minimized for a drag coefficient value CD = 4.2619 *kg/m*. I don't expect manufacturers will post numbers like this anytime soon, but wouldn't it be fun if they did?

Take-Aways

- The drag coefficient is a convenient parameter for characterizing the speed of a hull.

- The drag coefficient can be measured in the field using the test procedure laid out in this chapter. You now know how to perform a test to do this!

Extra Credit: Governing Equation Solution and Data Reduction

When the input paddling force is zero, the governing equation reduces to a simple homogeneous first-order nonlinear ordinary differential equation:

$$-C_D v^2 = m \frac{dv}{dt}. \tag{11.3}$$

We can rewrite Equation (11.3) as

$$\frac{dv}{dt} = -\frac{C_D}{m} v^2. \tag{11.4}$$

We define a new variable z which is the inverse of the speed v,

$$z \equiv v^{-1} \tag{11.5}$$

so that the governing equation assumes the simple form

$$\frac{dz}{dt} = \frac{C_D}{m}. \tag{11.6}$$

We can directly integrate this derivative to solve for z:

$$z = \frac{C_D}{m} t + k, \tag{11.7}$$

where k is a constant of integration. In terms of the speed v, this becomes

$$v = \frac{1}{\dfrac{C_D}{m} t + k}. \tag{11.8}$$

Since at $t = 0$ (the start of the coasting phase) $v = V_0$, the integration constant equals $1/V_0$. Substituting into the expression for v and simplifying, the speed as a function of time is then

$$v(t) = \frac{V_0}{1 + \left(V_0 \dfrac{C_D}{m}\right)t}. \tag{11.9}$$

• • •

Values of the drag coefficient were varied in the model to minimize the normalized mean square error e between the test data and the model prediction. We express the mean square error as

$$e = \frac{1}{N}\sum_{i=1}^{N}(v_i - w_i)^2, \tag{11.10}$$

where w_i are the speed values predicted by the model, v_i are the measured speed data value, and N is the number of data points. Equation (11.10) reflects how for each i-th time increment, the difference between test data and the model's predicted speed was computed, then squared. These squared values were summed and divided by the number of data points.

Why are the differences squared? Because at some times, the test data will be greater than the model's predicted value, while at others, it will be less. Therefore, sometimes the difference will be positive and sometimes negative. If you add positive and negative numbers, they can cancel each other out, suggesting that the model fit is excellent when it isn't. However, the squares of the differences are always positive, and their sum will always be a positive number as well – the algorithm iterates on the value of the Drag Coefficient that minimizes this sum. We divide by the total number of data points – e.g., *normalize* the data – because some data sets have more or fewer data points than others.

Further Reading

Nicholas Caplan, "The Influence of Paddle Orientation on Boat Velocity in Canoeing," *International Journal of Sports Science and Engineering*, **3**(, pp. 131-139 (2009).

James Buckmann and Samuel Harris, "An Experimental Determination of the Drag Coefficient of a Mens 8+ Racing Shell," *SpringerPlus 2014*, **3**:512.

CHAPTER 12

Start Me Up

Introduction

I'm a dinosaur. I prefer to derive solutions to physics problems whenever I can rather than numerically simulate them. Even when I can't write down a solution using pencil and paper, I look for limiting cases in the governing equations. What happens when one variable is large, or small? What happens when the hull/paddler/water system approaches a steady-state? I've found that this approach provides insights across hull types, paddles, and paddling disciplines without resorting to simulation.

In graduate school I witnessed the introduction of numerical simulation into the undergraduate engineering curriculum. Some faculty members argued that students would no longer need to build prototypes or conduct experiments. All hardware would be digitally simulated. Like my dinosaur-hood, this too is an extreme. I found the students I taught and supervised benefited from hands-on experience with hardware, sensors, signal conditioning electronics, and data acquisition. They began developing this mysterious and valuable attribute my undergraduate professors called "engineering judgment." We appreciate the limitations of simulation – and our incomplete understanding of calculus – when we witness integrator wind-up firsthand.[68]

---

68 Integrator wind-up happens when you digitally integrate (sum) a signal. For example, we know that acceleration is the derivative of velocity, consequently velocity is the integral of acceleration. They are inverse operations. Many newbies then conclude that you can integrate an accelerometer signal, and presto! You've got velocity. But suppose the accelerometer, its signal conditioning electronics, or your digitizer's front end

In this chapter, we'll dip our toes into simulation while keeping our heads out of the digital clouds and firmly seated in our hulls. Our model problem is the solo canoe whose drag coefficient we computed in Chapter 11 from experimental data. We'll consider various paddling force profiles to determine their impact on how this canoe – or rather, this canoe model – gets off a race starting line. Which is better: short choppy paddling strokes or long strokes? Does increasing paddle force over the first few strokes get the hull up to speed faster?

The Governing Equation

We'll use Newton's 2nd Law of Motion to derive a mathematical model that relates the paddling propulsive force F_p to the drag force and the hull speed v. Per Newton, the sum of all forces acting on the hull equals the total mass of hull plus paddler times the hull's acceleration. Including the drag force, which we previously showed was proportional to speed squared, the 2nd Law yields

$$F_p(t) = ma(t) + C_D v^2(t). \tag{12.1}$$

In this equation, m represents the mass of the hull plus paddlers and gear, while C_D is the hull's dimensional drag coefficient. The drag coefficient is specific to the hull's shape; different hulls have different drag coefficients. The dependent variable a represents the rate of change of the hull's velocity, otherwise known as its acceleration. Note that the propulsive force, acceleration, and velocity depend on time.

Equation (12.1) is called a *governing equation*. Solve it, and you can predict the hull's speed v as a function of time t given the drag coefficient C_D and the propulsive force F_p. Note that this is a simplified model of the moving hull. For example, it does not address the hydrodynamic effects of roll, pitch, or yaw, which all impact performance. Keep this in mind as we interpret solutions to the governing equation. Try not to read things into the model that aren't there!

The governing equation looks unassuming enough. However, owing to the velocity squared term, it's a nonlinear differential equation. Equation (12.1) is a *Ricatti equation* for those of you keeping score at home. For most propulsive force profiles F_p, the Ricatti equation does not lend itself to a closed-form solution; you can't just write down the answer. Bummer!

Our governing equation does, however, admit numerical solutions. To compute one, we'll introduce a teeny bit of calculus. The hull's acceleration is the time rate of change of its velocity, otherwise known as the velocity's time derivative. We can express acceleration as

$$a(t) = \frac{dv(t)}{dt}. \tag{12.2}$$

The notation d/dt means "rate of change with respect to time." This notation lets us re-write the governing equation as

---

have a constant ("DC") offset. In that case, you get to observe the time integral of a constant. Over time that integral trends toward infinity. Practically, the signal will grow until it exceeds some component's dynamic range and saturates the electronics or the signal's digital representation. Fun times!

$$\frac{dv(t)}{dt} = \frac{-C_D}{m}v^2(t) + \frac{1}{m}F_p(t)^{\cdot} \qquad (12.3)$$

Knowing the initial speed v_0 at a start time $t = 0$ makes this an *initial value problem*. We can compute an iterative numerical solution using the propulsive force as a function of time and the two constants.

Simulation Model

We'll assume that the hull's initial speed is zero, and use the drag coefficient for a solo hull computed from the field tests presented in Chapter 11. The solo canoe comprises a mass of 109 kg for the combined hull, paddler, and gear. The propulsive force is represented using equations developed in Chapter 6. As a baseline, we'll use a 1-second stroke duration. This corresponds to a 60 strokes/minute cadence. We'll initially assume the stroke has a 70% duty factor. This means the stroke's power phase lasts 70% of the stroke duration, or 0.7 seconds – see Chapter 7 for more detail about the paddling cycle, duty factor, etc.

All simulations were conducted using MATLAB. We used MATLAB's 'ode45' routine, which employs an explicit 4th-order Runge-Kutta method to solve the initial value problem. The baseline time range extended from 0 to 60 seconds in 1-millisecond increments. The ode45 solver internally adapts the time step as needed to address tolerance and solution convergence.

Results

First, let's compare the experimental hull speed data from Chapter 11 with our simulation model. As you'll recall, we brought the C-1 up to speed from a standing start, then coasted. This lets us use the governing equation's homogeneous solution to compute the hull's drag coefficient. You can compare experiment and theory in Fig. 12.1's two plots.

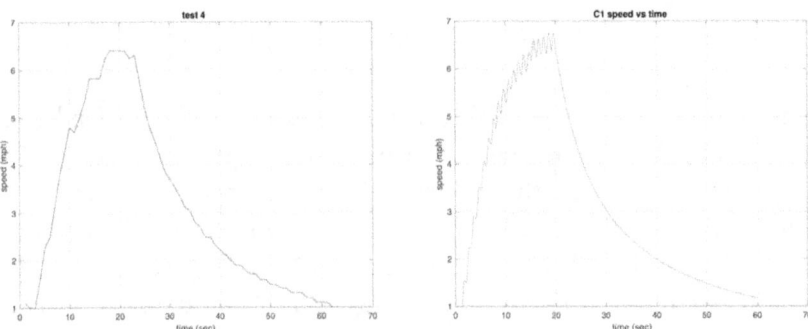

Figure 12.1: C-1 experiment and simulation.

The GPS used in the field digitized speed data at one sample per second. Consequently, the finer details of hull motion are lost, and the data at lower speeds is a bit chunky. However, the simulation reflects the general trends seen in the experimental data. The average speed rises as it would for an overdamped system, then drops inversely proportional to time during the coast. There was no way to measure propulsive force during the field tests. So we chose a peak force amplitude for the simulation that matched as best we could the steady-state hull speed just before the coast phase.

This simulation data is plotted by itself in Fig. 12.2. The quadratic drag force increases as the simulated hull approaches its cruising speed. The "hull" speed surges are rapidly damped by this drag. The drag force is lower at lower speeds. Consequently, the speed surges there don't damp out as quickly. Instead, the hull appears to "ratchet" up the drag curve at low speeds. If the simulated propulsive force isn't turned off at $t = 20$ seconds, the "hull" continues along at its average cruising speed. This is not surprising – but it is boring. As you'll see, we learn more about the system from its transient response.

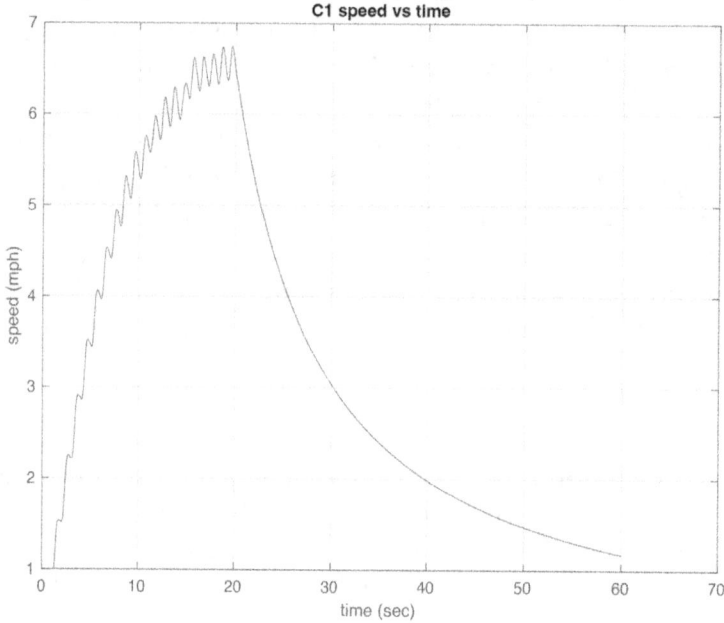

Figure 12.2: C-1 simulation data.

Now I've seen top-end racers with vastly different paddle strokes. A friend of mine (who shall remain nameless) has a fast, choppy stroke, while I know another expert paddler who employs longer strokes. Yet they've finished neck-and-neck at numerous races. So, I investigated two significantly different propulsive force profiles to make a point. The first, our baseline, has a one-second stroke duration and a 70% duty factor. The second has a two-second stroke duration (30 strokes/mininute cadence) and the same 70% duty factor. These two propulsive force profiles are plotted in Fig. 12.3. The goal was to determine whether a longer stroke, with all other factors held constant, might impact speed over time. The simulated hull speed curves for these two force profiles are plotted in Fig. 12.4.

As you can see in Fig. 12.4, the results are scarcely different. The lower cadence data shows speed oscillations at half the rate of the faster cadence. But aside from that, the average speed trend through these two cases is essentially identical.

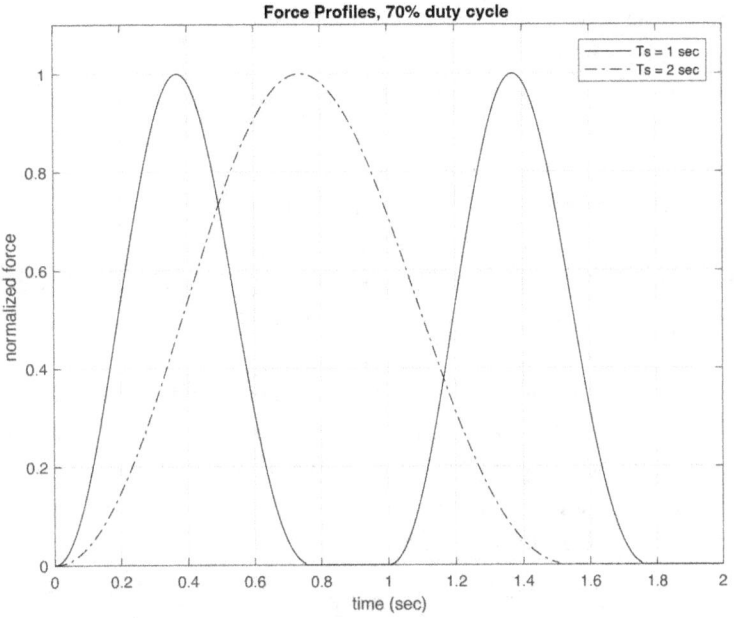

Figure 12.3: Normalized propulsive force profiles for two cadences.

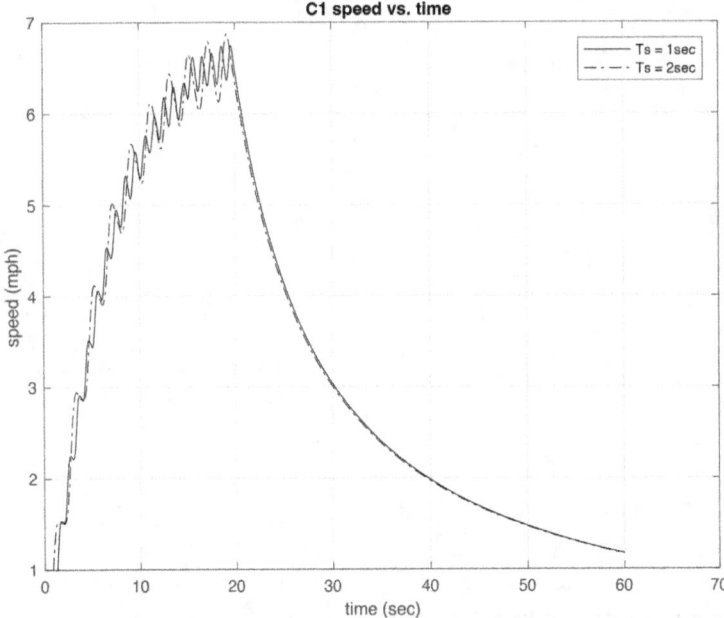

Figure 12.4: C-1 simulation data for fast and slow cadence force profiles, 70% duty factor.

Why does this happen? Recall from Chapter 7 that a hull's momentum – and thus its speed – is driven by the impulse the paddle imparts. Impulse is proportional to the area under the force versus time curve. The two force profiles in Fig. 12.3, corresponding to 60 and 30 strokes/mininute cadences, have identical areas under their curves. This is because they both use the same model of the force curve's shape and have the same duty factor and peak force. Thus, their impulses are the same.

There are two notable results from these simulations. First, it doesn't matter whether you have a fast or slow stroke if you apply the same impulse. Your body may be better suited to one cadence or another, so work with that. And second, Figs. 12.3 and 12.4 show that a model derived from Newton's 2nd Law and an analysis developed using conservation of momentum yield the same result. Since we're operating in the world of classical mechanics, this had better be the case! But if you're new to physics, you might wonder why I might choose one method of attack versus another. The reason? Convenience. They both describe the same reality, as demonstrated by our results.

So, if stroke cadence won't get a C-1 up to speed faster, what will? We next looked at the role duty cycle plays. Using our baseline paddling model with a 60 strokes/min cadence, we compared 60%, 70%, and 80% duty factors. Recall that a higher duty factor means a faster recovery; the stroke's power phase is longer as a percentage of the stroke duration. These three force profiles are plotted in Fig. 12.5.

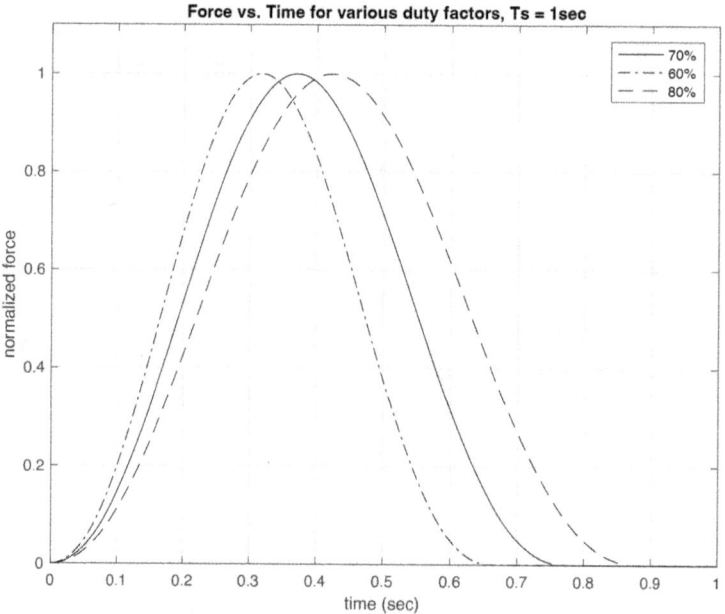

Figure 12.5: Normalized propulsive force profiles for three duty factors.

The simulation results for these three propulsive force inputs are plotted in Fig. 12.6. The higher duty cycle force profile yields a greater top speed before the coast; the lower duty cycle profile yields a lower top speed. This isn't surprising when we consider the impulse imparted by each profile. Since the stroke duration is the same in all three cases, the only difference is the duty cycle. For a fixed cadence, a higher duty cycle provides more impulse. This yields greater hull momentum and thus speed. Yay – physics works! But notice that all three cases come up to speed at about the same time. A higher duty cycle yields a greater top end. And that top end occurs at about the same time for each.

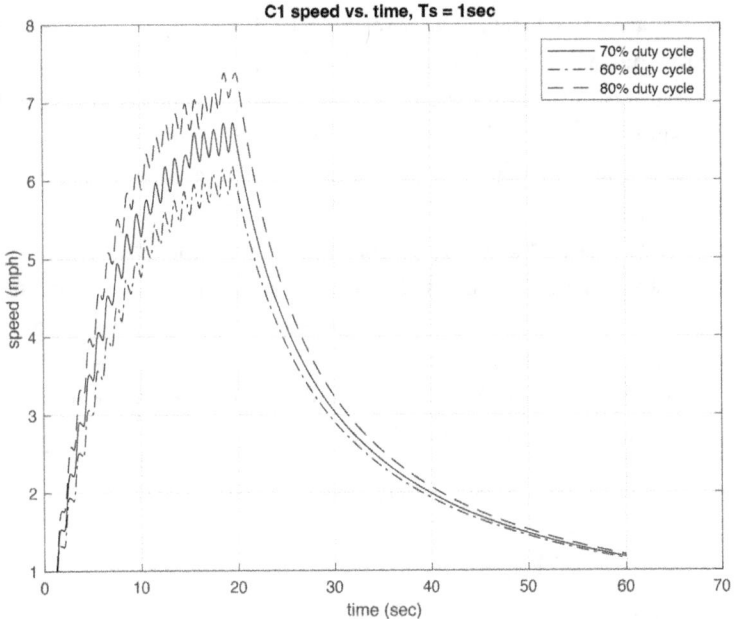

Figure 12.6: C-1 simulation data for force profiles having various duty factors.

What can we do to get a hull up to speed faster? Many of us apply greater force over our first few strokes. Our propulsive force tapers over several strokes, after which we engage our steady-state paddling mechanics. To capture this, we simulated a force profile with varying peak force over the first ten strokes. This is depicted in Fig. 12.7. The first stroke has

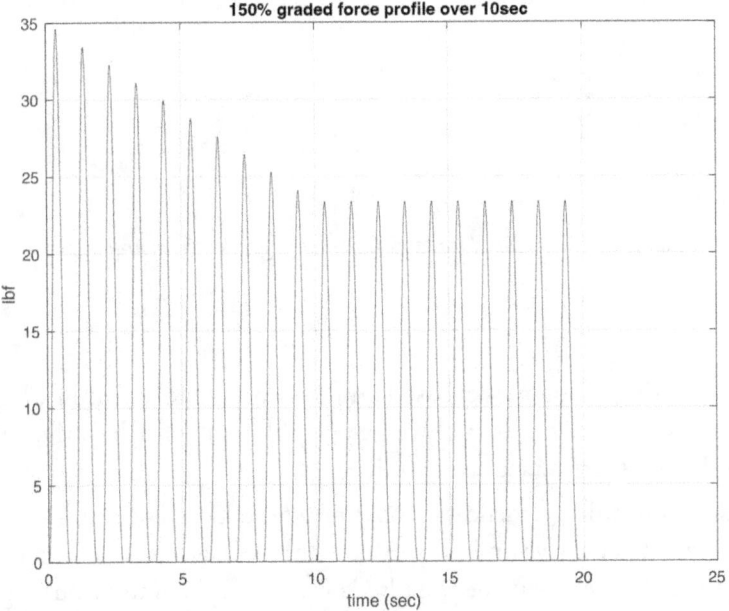

Figure 12.7: Variable propulsive force profile.

150% of the steady-state peak force. The force profiles for each stroke taper linearly until t = 10 seconds. After that, they are identical to the T_s = 1 second stroke plotted in Fig. 12.3.

It should come as no surprise that harder initial strokes bring the simulated C-1 up to speed faster, as shown in Fig. 12.8. The graded profile's stronger initial strokes get the simulated hull "out of the hole" faster, ratcheting up the drag curve more quickly despite having the same cadence.

You can combine effects and grade both the stroke force and stroke duration over time, e.g., initially slower and stronger strokes transitioning to a faster steady-state cadence and peak force. That result is left for the reader to investigate. But considering what we know about physics, we can predict what the result will be. If the force profile entails greater impulse, you'll go faster. If it entails greater impulse over a group of strokes than others, you'll speed up more quickly.

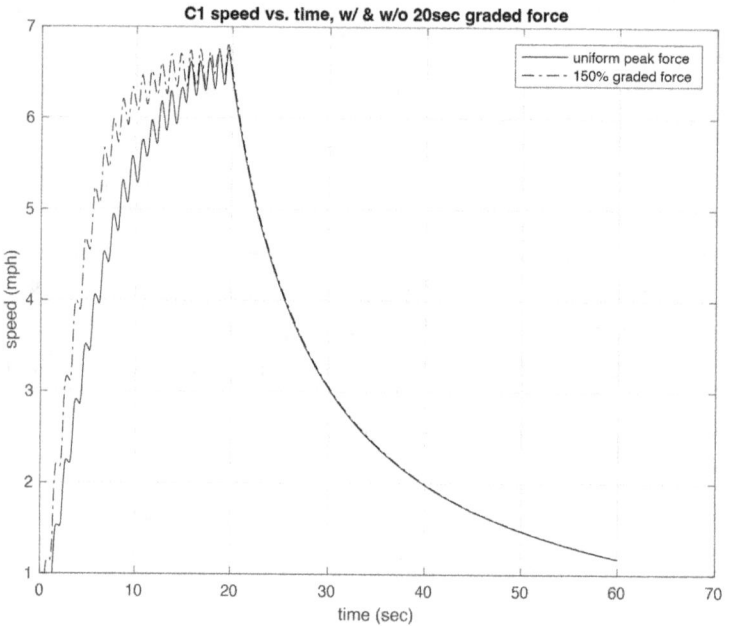

Figure 12.8: C-1 simulation data for uniform and graded force profiles, Ts = 1sec.

Take-Aways

- We developed a simulation model based on the solo canoe studied in Chapter 11. This lets us determine how various propulsive force profiles influence how the hull – or rather, the hull model – gets off a race starting line.

- Faster and slower paddling cadences do not impact the simulated hull's speed versus time curve for a given force curve and duty cycle. This reflects the role impulse plays – the area under the force curve. If the impulse is the same, the hull's resulting momentum and thus speed will be the same.

- Changing the stroke's duty factor – the percentage of the stroke where the paddle provides propulsion – changes the simulated hull's top end speed. A greater duty factor means greater impulse.

- You get off the line faster by employing more powerful strokes at the start for a given stroke rate.
- Whether you're a hummingbird or a grinder, you now know that either approach can be equally effective. Pick whichever style suits your biomechanics and neuromuscular capabilities.

Extra Credit – the Runge-Kutta method

The Runge-Kutta method for numerically solving first-order ordinary differential equations is based on the solution's Taylor Series expansion. From freshman calculus, if the velocity $v(t)$ is smooth, it can be expressed at time $t + h$ in terms of its derivatives,

$$v(t+h) = v(t) + h\frac{dv(t)}{dt} + \frac{h^2}{2!}\frac{d^2v(t)}{dt^2} + \ldots \quad (12.4)$$

If we can calculate the velocity's higher-order derivatives through order p, then the value of v at a discrete moment in time $n+1$ can be derived in terms of velocity and velocity derivative values at the previous time point n,

$$v_{n+1} = v_n + h\frac{dv_n}{dt} + \frac{h^2}{2!}\frac{d^2v_n}{dt^2} + \frac{h^3}{3!}\frac{d^3v_n}{dt^3} + \ldots + \frac{h^p}{p!}\frac{d^pv_n}{dt^p}. \quad (12.5)$$

Staring at time point n, we compute the various terms in Equation (12.5)'s right-hand side to approximate the velocity at time point $n+1$. Then we compute the right-hand side at time point $n+1$ to approximate the velocity at time point $n+2$. Assuming everything converges, the process continues iteratively until the velocity is computed for all desired time points.

The Runge-Kutta method approximates this Taylor Series expansion without computing the higher-order derivatives. If we re-write Equation (12.3) very generally as

$$\frac{dv}{dt} = g(v,t) \quad (12.6)$$

then the Runge-Kutta approximation of Equation (12.5) is

$$v_{n+1} = v_n + \frac{1}{6}(k_0 + 2k_1 + 2k_2 + k_3) \quad (12.7)$$

for a time step of length h, and

$$k_0 = hg(v_n, t_n) \qquad \text{(12.8)}$$

$$k_1 = hg\left(v_n + \frac{1}{2}k_0, t_n + \frac{h}{2}\right)$$

$$k_2 = hg\left(v_n + \frac{1}{2}k_1, t_n + \frac{h}{2}\right)$$

$$k_3 = hg(v_n + k_2, t_n + h)$$

As with the discretized Taylor series, staring at time point n, we compute the various terms in Equation (12.7)'s right-hand side to approximate the velocity at time point $n+1$. The process continues iteratively until we obtain the velocity for all desired times. As the time step size h becomes infinitesimally small, the Runge-Kutta method converges to the Taylor Series representation through terms of order h^4.

Further Reading

F. Acton, *Numerical Methods That (Usually) Work*, Harper & Rowe, New York NY, pp. 129-156 (1970).

G. Forsythe, M. Malcolm, and C. Moler, *Computer Methods for Mathematical Computation*, Prentice-Hall, Englewood Cliffs NJ, pp. 121-146 (1977).

CHAPTER 13

Cutting Corners

Introduction

When is it advantageous to take the inside track in a shallow water turn during a race? This simple question is far more complex than it may appear.

A precise answer requires understanding the assumptions we make in posing the question. Often, when constructing a model or performing analysis, we're not even aware that we've made assumptions! You likely envisioned your hull and a competitor's entering the turn simultaneously. But are these hulls hydrodynamically the same? Does one hull behave differently than the other when turned in progressively tighter arcs? Or does it begin to carve the turn and thus change our assumptions about its hydrodynamics? If the turn is "shallow" for one hull, is the other hull in "deep" water, or something else? Fundamentally, how do we define shallow and deep? How does the water depth vary across the turn? Is the turn circular, elliptical, or something else? Is there current, and if so, is the current uniform across the turn? Do the paddlers in each hull have the same biomechanical efficiency? Or are we only concerned with hydrodynamics?

Each of these factors introduces its own complications. We can address many of them via numerical simulation. However, the resulting predictions may only apply to a particular turn, on a particular river, and to particular hulls and paddlers. Where's the fun in that?

We'll instead consider a more general approach to the corner-cutting problem by introducing several simplifying assumptions. Our assumptions influence the problem formulation, yielding a tractable model whose analysis provides insight into route selection. As we'll see, these assumptions must always be kept in mind since broadly interpreting an already constrained problem can lead to mistaken conclusions. Think of this chapter as an object lesson in making and tracking assumptions – science and engineering's version of eating your vegetables. As you'll see, the general problem of route selection is multifaceted and seldom lends itself to detailed analysis.

Assumptions

To start, we'll assume that the two turning hulls are hydrodynamically the same throughout, with the same trim. Consequently, their friction and form drag are the same. We'll define shallow and deep water using Chapter 3's analysis. Shallow water wave drag won't impact either hull unless the water's depth is less than half the hull's length. And we'll assume that each hull moves at the surface wave phase speed corresponding to the local water depth; see Chapter 3 for more detail on phase speed and hull speed. With identical hulls, moving at the local surface wave speed ensures comparable wave drag for each hull.

So now we'll introduce the following assumptions:

- All paddlers provide equal propulsive power.
- One hull takes an inside route through shallow(er) water, and the other takes an outside route through deep water. By "deep" we mean the depth is greater than half the outside hull's waterline length.
- The water depth increases monotonically from the turn's inside to its outside. This is an implicit assumption; otherwise, interpreting the results gets weird. (We can pose the problem with the depth decreasing toward the turn's outside. But that, as they say, is left as an exercise to the reader.)
- The turn is circular.
- Within the turn, the water depth stays constant along any circumferential arc.
- We'll restrict ourselves to the linear (small wave) approximation employed in Chapter 3. This precludes soliton waves, breaking, or other nonlinear wave effects.
- Each hull moves at a constant speed through the turn. We'll neglect all acceleration and deceleration effects from entering and exiting shallow(er) or deep(er) water.
- We address added current in the Extra Credit section. For the moment, we'll assume that both hulls are in a reference frame moving with the flow, and the flow is uniform cross-stream. In both hulls' reference frame, the current is zero.

Model Formulation

Consistent with the assumptions above, consider the turn depicted in Fig. 13.1. Two hulls running abreast simultaneously enter a 90-degree counter-clockwise turn. The first follows a circumferential arc of length P_1 at a radius R_1 over an angle θ ranging from 0 to 90 degrees.[69]

---

69 As we'll see, the curve's angular sweep can assume any value consistent with the stated assumptions. A ninety-degree turn is a convenient example.

The second follows an arc of length P_2 at a radius R_2 over the same angle θ. We assume that the water at radius R_2 and beyond is deep throughout the turn. The subtended angles are

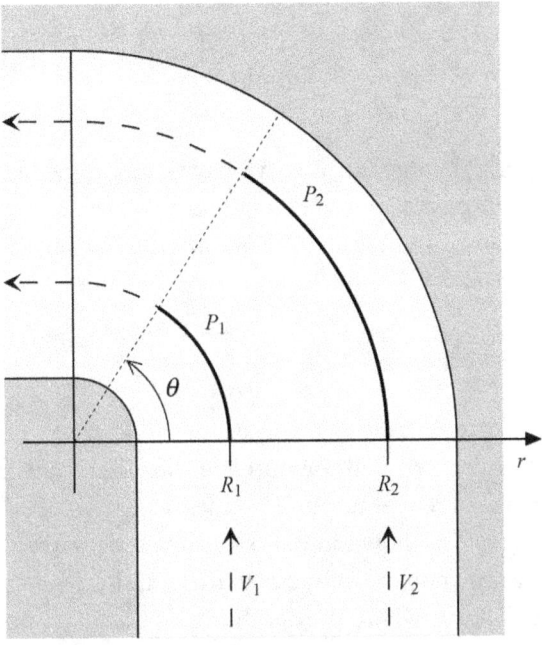

Figure 13.1: Turn geometry, plan view.

equal since we wish to determine how small a radius R_1 the inner hull can follow and keep up with the hull in deep water.

In Chapter 3, we defined a nondimensional Froude Number[70] F_r as the ratio of hull speed V to the surface water wave phase speed c_p:

$$F_r \equiv \frac{V}{c_p}.$$ (13.1)

When the hull speed is greater than the phase speed, the Froude Number F_r is greater than one. In that case, you're paddling faster than the water's surface "wants" to move, and wave drag increases. So, we'll introduce another assumption: Both the "inner" and the "outer" hulls move at the surface wave phase speed corresponding to their respective water depths. In other words, $F_r = 1$ for both hulls. This is one way of stating that the paddling effort in overcoming wave resistance is the same for both hulls.[71]

Given these two hull speeds – really, two surface wave phase speeds since each hull's

---

70 Note that we're specifying *a* Froude Number. You can define the Froude Number in other ways, just like you can specify the Reynolds Number in various ways. You'll be fine if you clearly and consistently define the quantities used to compute either quantity.

71 This further assumes that the paddler(s) in each hull has the same paddling biomechanical efficiency. As you can see, these assumptions pile up!

Froude Number is assumed to equal 1 – the problem reduces to finding the radius R_1 for a given inner hull water depth D that allows the two hulls to keep pace with each other through the turn. For our hull in shallow(er) water, the phase speed is

$$c_p = \sqrt{\frac{gL}{2\pi}\tanh\left(\frac{2\pi D(r)}{L}\right)}, \tag{13.2}$$

where g is the gravitational constant, and we have substituted the hull length L for the surface wavelength; the hull is thus traveling at its hull speed we discussed in Chapter 3. The depth D is a function of the radius r. For our hull in deep water, the expression for phase speed simplifies to

$$c_p = \sqrt{\frac{gL}{2\pi}}, \tag{13.3}$$

where we have once again substituted the hull length L for the surface wavelength.

We'll now consider a concrete example: Two identical hulls, 18.5' long, moving in a turn having bottom bathymetry[72] as shown in Fig. 13.2. The outer radius R_2 corresponds to deep water, where "deep" means more than about half the hull's length L. Consequently, the allowable outer radius R_2 starts where the depth equals $L/2$ and ranges outward from there. This is consistent with the deep-water assumption since the depth there exceeds $L/2$. The corresponding phase speeds are plotted in Fig. 13.3.

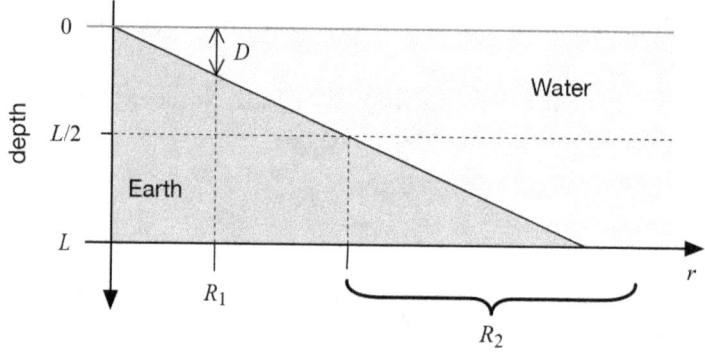

Figure 13.2: Turn bathymetry.

From high school physics – or perhaps high school algebra – we know that the distance you travel equals speed multiplied by moving time. Fig. 13.3 shows that the outer hull always moves at the deep-water phase speed for our model problem. Its shortest route then is the radius where the depth begins decreasing, at 9.25'. The inner hull, paddling at the local surface wave phase speed, moves progressively slower with decreasing depth. The inner

72 Bathymetry is a fun word for underwater topography.

hull must follow a tighter arc with less circumferential distance to keep up. The solid line plotted in Fig. 13.3 defines this "keep pace" radius.

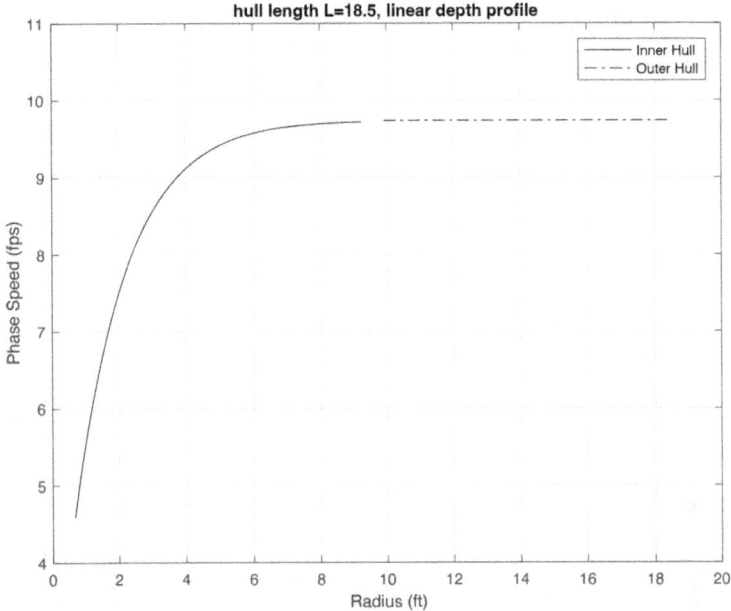

Figure 13.3: Hull speed vs. radius, 18.5' hull, linear depth profile.

We can extend our result to other depth profiles and hull lengths by nondimensionalizing both the depth D and the inner turn radius. This process is also called *normalizing*. A normalized quantity is a ratio. When this ratio is small, the numerator is small compared to its denominator, and vice versa when the ratio is large. Both "large" and "small" are concepts that imply large and small compared to something. By carefully selecting the something, this comparison is automatically built into the normalized quantity.

We normalize the depth D by dividing it by half the hull length $L/2$. This depth is physically significant because $L/2$ defines the transition between deep and shallow water. The resulting nondimensional depth indicates whether the depth is small or large compared to this breakpoint. When the normalized depth equals one, the inner hull is traveling at the $L/2$ depth. Then, we normalize the inner hull radius R_1 by dividing it by the deep-water hull's radius R_2. A normalized radius of 0.75 means that if the hull in deep water follows a circular route with a 16' radius, the inside hull follows a circular route with a 12' radius. This is essentially the radius ratio defined in Equation (13.6) of the Extra Credit section. The resulting plot appears in Fig. 13.4.

The plot describes the ratio of inner and outer turn radii – whatever the outer turn radius is – where the inner hull can precisely keep pace given the depth of water under it. The plotted curve has an interesting property. Anywhere on the plot below the curve means the inner hull exits the turn in the lead. A straightforward example is when the normalized depth equals one. At this normalized depth both hulls are in deep water. Consequently, the shorter route wins - and all the shorter routes where the inner radius is less than the outer radius lie below the plotted curve. It's just that as the water gets shallow(er), wave drag slows you down, so you must choose a shorter route to compensate – or paddler harder! In essence, the curve represents the maximum turn radius the inner hull should follow for a given depth if its paddlers don't want to lose ground. Below it, the inner hull exits first for the given shallow(er) depth.

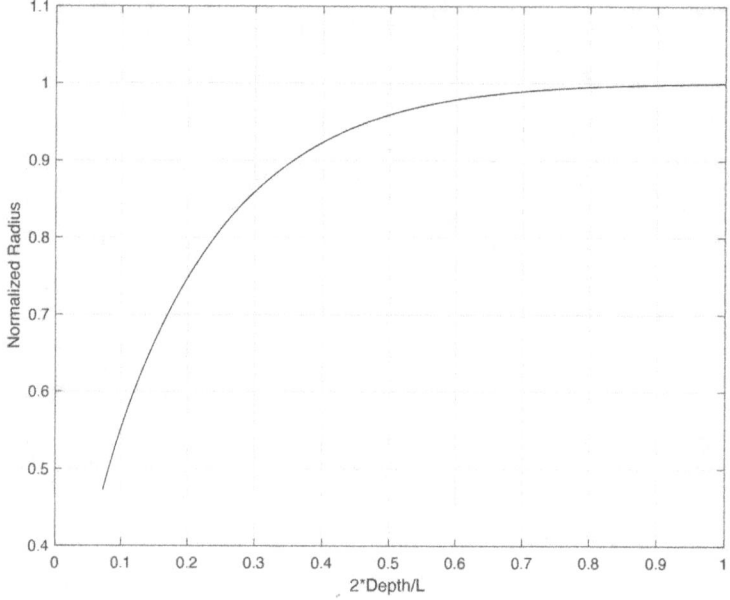

Figure 13.4: Normalized radius R1/R2 vs. normalized depth.

Interpreting Fig. 13.4 can be problematic when we consider the area above the plotted curve. When the ordinate equals one, the inner and outer hulls are traveling along the same radius. In that case, what does it mean to be "inner" or "outer"? For a normalized depth of one, ordinate values greater than one mean the inner hull travels outside the outer hull. You might excuse this by saying, "Well, both hulls are in deep water, and the one we labeled inner is merely outside the other and won't be able to keep pace with it."

But things get strange for normalized depths less than one. Consider the area above the curve for any smaller normalized depth value, such as 0.3. A normalized depth of 0.3 implies shallow water. We can find a normalized radius of one on the plot for this depth, which implies that both hulls are traveling along the same radius, but one hull is in shallow water while the other is in deep water. This is, of course nonsensical. As noted in the chapter introduction, the assumptions inherent in the analysis must inform our interpretation of the results. This is good advice for interpreting Fig. 13.4, as well as any figure or conclusion in a technical article or book.

Take-Aways

- In any analysis, assumptions must be carefully laid out and adhered to. Results must be interpreted in light of the assumptions, too.
- Cutting a circular shallow water corner saves time if the water isn't "too shallow," subject to the assumptions and limitations of our analysis.
- We derived a turn radius criteria using the depth-dependent surface wave phase speed. Our analysis shed light on the why of cutting corners.
- Unfortunately, the paddling radius that lets you keep pace with a hull in deep water depends on the square root of the depth's hyperbolic tangent. To compute this, you need to know the depth as a function of the turn radius. You should probably also bring a calculator! Or pack a copy of Fig. 13.4.
- Sometimes the simplifying assumptions needed to undertake an analysis limit the range of application so much that we might question why we bothered trying. A more balanced view is that some problems have greater complexity than we assumed. The turning problem is a prime example of hidden complexity.

Extra Credit: Going Deeper (and Shallower)

We can approach the corner-cutting problem by computing how far each hull travels over a time interval T. We then solve for the respective radii that ensure the two hulls cover the same angle θ and thus stay abreast of each other. For the inner hull, consistent with our assumptions, the circumferential path length P_1 equals the surface wave phase speed multiplied by our moving time,

$$P_1 = \theta R_1 = \sqrt{\frac{gL}{2\pi} \tanh\left(\frac{2\pi D(r)}{L}\right)} \cdot T \text{,} \tag{13.4}$$

We use Chapter 3's expression for the surface wave phase speed c_p as the hull speed, substituting the hull length L for the wavelength. The depth D is a function of the radius r; this dependence will be an essential consideration in a moment.

The hyperbolic tangent displays a sigmoid-like behavior as a function of its argument, with asymptotes at +/- 1, shown in Fig. 13.5 for an 18.5' long hull and various depths D. As you can see in the plot, for values of the argument with magnitude[73] greater than about 3, tanh(•) can be approximated by the constant 1. This corresponds to deep water for that hull. Consequently, for the outer (deep water) hull, the circumferential path length P_2 equals the surface wave phase speed for deep water times the moving time:

$$P_2 = \theta R_2 = \sqrt{\frac{gL}{2\pi}} \cdot T \text{.} \tag{13.5}$$

---

73 We're only concerned with positive values of the depth; who knows what a negative depth is? The hyperbolic tangent function itself allows both positive and negative arguments. Plus, the plot is pretty.

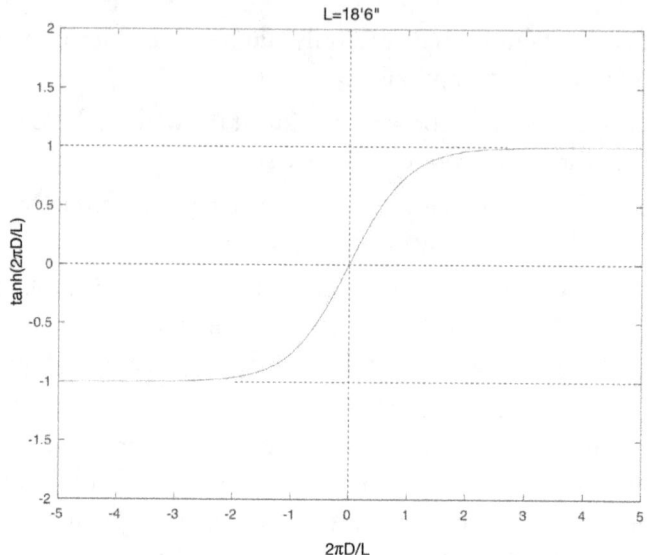

Figure 13.5: Hyperbolic tangent function.

Since the hulls are assumed to paddle abreast, the angle θ is the same in both hulls. The time T is the same as well since the hulls are assumed to enter and exit the turn at simultaneously. Solving both expressions for the angle θ, and since things equal to the same thing equal each other,

$$\frac{R_1}{R_2} = \frac{\sqrt{\frac{gL}{2\pi}\tanh\left(\frac{2\pi D(r)}{L}\right)}}{\sqrt{\frac{gL}{2\pi}}}. \tag{13.6}$$

Equation (13.6) defines the inner radius R_1 in shallow(er) water which allows you to keep pace with a hull further offshore traveling in deep water and is plotted in Fig. 13.4 for a (somewhat) arbitrary depth profile. In Equation (13.6), the radius R_1 is the radius at which the depth equals D. We cannot compute the surface wave phase speed for a given shallow water depth D – the numerator on the right-hand side of Equation (13.6) – then vary the radius R_1 independently. Since the depth D is a function of the radius the problem is recursive.[74] Further, the radius R_2 must correspond to "deep water," and be greater than R_1.

If the hyperbolic tangent function makes you feel all squidgy, you can use the shallow water approximation for the inner hull's surface wave phase speed. The shallow water approximation is derived using the Taylor series[75] expansion of the hyperbolic tangent function, vis

---

74 If you solve the equation for the inner radius R_1, you define this radius in terms of itself since depth D depends on R_1.

75 Technically this is a Maclaurin series since we expand the function about at x rather than $(x - a)$. But Taylor gets all the credit.

$$\tanh\left(2\pi D/L\right) = \left(2\pi D/L\right) - \frac{\left(2\pi D/L\right)^3}{3} + \frac{2\left(2\pi D/L\right)^5}{15} - \dots \qquad (13.7)$$

We'll now consider small values of the argument $2\pi D/L$, implying that the depth D is small compared to the hull length L.[76] When the hyperbolic tangent's argument is much less than one, the next term in the expansion will be smaller in magnitude than the previous one. So, we retain only the first term in the series for truly small depth values compared to hull length. The higher terms can be neglected. This approximation is illustrated in Fig. 13.6, where the exact form of the hyperbolic tangent is plotted along with the simple linear approximation for an 18'6" long hull. Interestingly, the approximation is best only for water that is about 1.5' deep or less. We'll revisit this limitation in a minute.

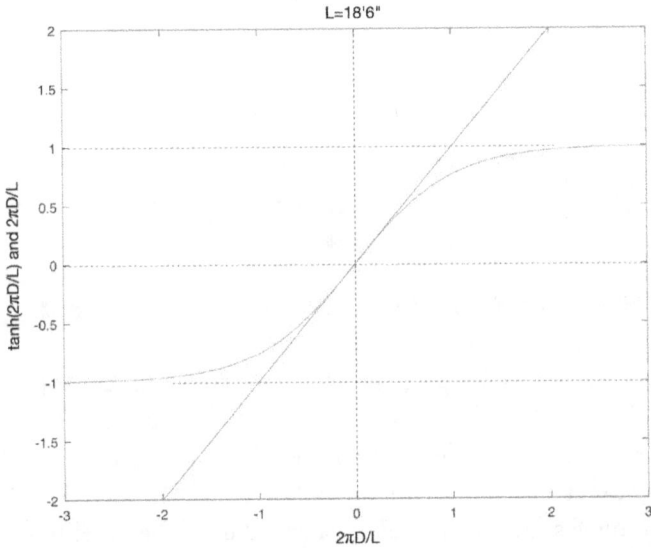

Figure 13.6: Hyperbolic tangent and a linear approximation.

Consequently, for an inner hull traveling in very shallow water, you obtain the more straight-forward expression for the two radii,

$$\frac{R_1}{R_2} \approx \sqrt{\frac{2\pi D(r)}{L}} \quad \text{for} \quad \frac{2\pi D(r)}{L} \ll 1. \qquad (13.8)$$

So, for cutting *very* shallow water corners consistent with the assumptions above, the turn radius that lets you keep pace with a hull traveling in deep water varies as the square root of depth. This ratio is plotted in Fig. 13.7 for the approximation above, and for the "exact" expression that uses the un-approximated hyperbolic tangent function. Fig. 13.7 shows that

---

76 As noted above, "small" always implies small compared to some other quantity.

the exact and approximate solutions diverge at about 1' of depth; above about 2.5' of depth, the approximation becomes nonsensical since it suggests the inner radius be greater than the outer radius (and thus contradicts one of our assumptions). Sorry, folks. If you felt squidgy about the hyperbolic tangent function, you'd best learn to embrace it.

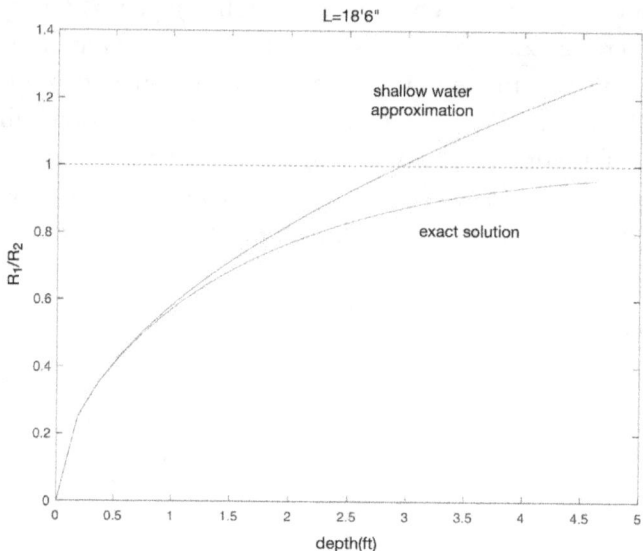

Figure 13.7: Radii ratio for the exact and approximate solutions.

You can use Fig. 13.7 to select an inner radius R_1 to cut the corner corresponding to a depth D given the outer radius R_2. Perhaps most interesting is what happens when the water depth approaches $4.5 - 5'$, at or just a little deeper than so-called "concrete water." In that case, the radii ratio (using the exact solution) asymptotically approaches one. At this $4.5 - 5'$ depth, the inner hull radius moves toward the deep-water radius of the outer hull. This is because both hulls are now in (effectively) deep water. Why do they become equal? Because we were solving for an inner radius that enabled the hulls to run alongside, not for an inner radius that let the inner hull emerge from the turn ahead.

Extra Credit: Current Effects

It's straight-forward to introduce a spanwise varying current into the above analysis. Suppose the inner hull travels in a constant current with speed V_1 throughout the turn. In that case the circumferential arc length it subtends over time T to keep abreast of the outer hull is a restatement of distance equals speed times time, where speed is now the sum of the phase speed and current speed:

$$P_1 = \theta R_1 = \left[\sqrt{\frac{gL}{2\pi} \tanh\left(\frac{2\pi D(r)}{L} \right)} + V_1 \right] \cdot T \qquad (13.9)$$

Similarly, for the outer hull

$$P_2 = \theta R_2 = \left[\sqrt{\frac{gL}{2\pi}} + V_2 \right] \cdot T \tag{13.10}$$

Solving for the angle θ, and equating things equal to the same thing yields

$$\frac{\sqrt{\dfrac{gL}{2\pi}} \tanh\left(\dfrac{2\pi D(r)}{L} \right) + V_1}{R_1} = \frac{\sqrt{\dfrac{gL}{2\pi}} + V_2}{R_2} \tag{13.11}$$

So, if you know the shallow water depth D at R_1 and the current speeds at the respective radii, you can compute the radius for "cutting the corner" that keeps you abreast of the hull in deep water. Or, for intermediate depths, you can use the exact expression.[77]

The primary limitation in using this result is obtaining the respective current speeds V_1 and V_2. There are innumerable references that address stream flow in aggregate (see Further Reading for examples). These publications provide models of sedimentation transport for a given streambed and flow rate, flow rate as a function of level, etc. But none provide cross-stream flow speed models. If you have current data, you can crunch the numbers. Otherwise, you're left to make assumptions that likely won't generalize across a set of rivers or even turn-to-turn for a given river.

Further Reading

Luna Leopold and Thomas Maddock, Jr., "The Hydraulic Geometry of Stream Channels and Some Physiographic Implications," Geologic Survey Professional Paper 252, United States Government Printing Office (1953).

Jud F. Kratzera, Daniel B. Hayes, and Bradley E. Thompson, "Methods for interpolating stream width, depth, and current velocity," *Ecological Modelling* **196**, pp. 256–264 (2006).

---

77 You can simplify either result a bit by using the difference in speed between the two hulls, defining a new speed V_3 as $V_2 - V_1$, and after that setting V_1 to zero; this means you've transformed the problem to an inner hull-centric coordinate frame.

CHAPTER 14

Speed Above Replacement

Introduction

Is there a way we can leverage science to determine how a paddler performs in comparison to others in the same boat? Is there something like baseball's Wins Above Replacement ('WAR') for paddling? If so, this tool could help fill out teams for outrigger or dragon boat racing or determine which seat best suits a C-4 paddler.

It turns out there's a straightforward metric we can compute from field measurements. The metric leverages what we learned about the relationship between hull speed and power in Chapter 1. We'll include the derivation in this chapter's main body since it relies on high school algebra. Besides, just presenting the result bypasses the fun of seeing the why.

Speed Versus Power in the Boat

As you may recall from Chapter 1, the speed of a paddled hull is proportional to the cube root of the applied paddling power. Most generally, for a hull having N paddlers, this is expressed as

$$V_N = c_N \sqrt[3]{\sum_{i=1}^{N} P_i} \, , \tag{14.1}$$

where V_N is the speed of the N-paddler hull. The constant c_N combines all the N-paddler hull's hydrodynamic qualities into a single constant. As we'll see, the specific value of c_N is irrelevant since we'll be computing ratios. The variable P_i is the power generated by each paddler, numbered from 1 through N. The large sigma in Equation (14.1) is shorthand for summation; we compute the total paddling power in the boat by summing each paddler's power contribution.

For a six-person outrigger ("OC-6"), we can write Equation (14.1) somewhat laboriously as

$$V_6 = c_6 \sqrt[3]{P_1 + P_2 + P_3 + P_4 + P_5 + P_6} \,, \tag{14.2}$$

and for a four-person canoe ("C4"), it can be written as

$$V_4 = c_4 \sqrt[3]{P_1 + P_2 + P_3 + P_4} \,. \tag{14.3}$$

First, let's consider the C4 case. Fig. 14.1 is a C4 in plan view, paddled from left to right. In the first case, paddlers in all four seats are paddling, represented by gray seats. We assume that the hull has been brought up to speed and is moving at a steady pace over flat water with no current. In the second case, the paddler in the 2nd seat has stopped paddling – hence the x-ed out seat – and is sitting upright. This paddler has also feathered their paddle to minimize air drag from their blade.

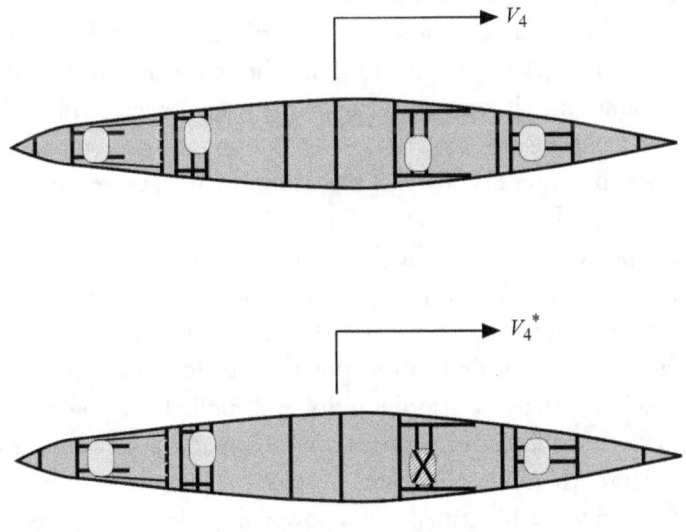

Figure 14.1: C4, no current.

The hull speed for the first case appears in Equation (14.3). For the second case,

$$V_4^* = c_4 \sqrt[3]{P_1 + P_3 + P_4} \,, \tag{14.4}$$

where V_4^* represents the hull speed with one paddler – here, the second from the bow – offline. In principle, this could represent any one of the paddlers 1 through 4 going offline. In practice, taking the stern paddler offline is problematic since they call switches, etc. The ratio of hull speeds for these two cases is

$$\frac{V_4}{V_4^*} = \frac{\sqrt[3]{P_1 + P_2 + P_3 + P_4}}{\sqrt[3]{P_1 + P_3 + P_4}} \cdot \qquad (14.5)$$

By taking a ratio, the hull's hydrodynamic coefficient c_4 cancels out. We are comparing paddlers in the same hull, not comparing hulls. To characterize paddler number 2 compared to their teammates, we assume that paddlers 1, 3, and 4 still exert the same paddling power, which we label P. We also assume that when all four paddlers are active, paddler number 2's power wasn't wildly different from this average power P; we wish to determine how one paddler's power impacts this average. Consistent with these assumptions, we quantify the impact of taking paddler number 2 offline via the speed ratio

$$\frac{V_4}{V_4^*} = \frac{\sqrt[3]{4P}}{\sqrt[3]{3P}} = \sqrt[3]{\frac{4}{3}} = 1.10064242 \qquad (14.6)$$

If paddler number 2 can apply the same average power as their teammates, then the ratio of hull speeds for these two cases should be about 1.1. In other words, the C4 moves 10% faster when paddler number 2 is paddling vs. when they are not yet are still in the boat.

If the speed ratio is greater than 1.1, paddler 2 provided more power than the average. Why is this so? First, Equation (14.6) assumes *a priori* that all paddlers' power is the same; we're just looking for changes to this ideal ratio. The numerator in Equations (14.5) and (14.6) is always the same since it represents the team's total power. Only the denominators can change in these equations.

There are four possible ways to construct the denominators, corresponding to each paddler going offline in turn. This yields four different three-paddler groupings. If all paddlers were equal, the computed ratio would be the same in every case. If they are not, and we take a more powerful paddler offline, the remaining paddler power we sum to compute the denominator is smaller than the other cases. When the denominator becomes smaller, the ratio exceeds 1.1, and we've identified a more powerful paddler than the average.

If the speed ratio is less than 1.1, paddler number 2 applied less power than the average. By taking a less powerful paddler offline, the remaining paddler power we sum to compute the denominator is larger than the other cases. When the denominator becomes larger, the ratio dips below 1.1, and we've identified a less powerful paddler than the average.

We'll call our ratio the C4 paddler Speed Above Replacement ("SAR") threshold *for that complement of paddlers*. It is not a valid comparison for all paddlers in a paddling club, a country, or the world since every paddler cannot be in the boat during our test Such a far-ranging comparison requires measuring all paddlers in the same hull at the same time. That's neither practical nor geometrically possible in most cases!

We can similarly compute the Speed Above Replacement for an OC-6:

$$\frac{V_6}{V_6^*} = \frac{\sqrt[3]{P_1 + P_2 + P_3 + P_4 + P_5 + P_6}}{\sqrt[3]{P_1 + P_3 + P_4 + P_5 + P_6}}, \qquad (14.7)$$

which for the uniform baseline power assumption reduces to

$$\frac{V_6}{V_6^*} = \frac{\sqrt[3]{6P}}{\sqrt[3]{5P}} = \sqrt[3]{\frac{6}{5}} = 1.06265857 \tag{14.8}$$

This is the OC-6 SAR threshold for a paddler in comparison to their teammates: 1.063.

You may begin to see a pattern here. As the number of paddlers in a boat increases beyond six, this ratio gets smaller. The speeds we measure in the field for computing SAR are prone to measurement error. As the number of paddlers N increases, the SAR threshold ratio asymptotically approaches one. Consequently, for a large number of paddlers – like in some dragon boats – the speed reduction from taking one paddler offline falls below the sensor measurement error.

Practical Considerations

We must account for several real-world factors when measuring SAR during a field test. First, taking a paddler offline means that the hull speed must stabilize before gathering the speed data used in our computation. As we saw in Chapter 11, it takes nearly a minute for a C-1's speed to stabilize (i.e., drop to zero) after taking its paddler offline. An on-water assessment must include a time interval for accelerating to a uniform baseline speed. And it must include a stabilization time for the decelerating hull to settle at its new speed.

You must average the two speeds comprising the SAR calculation to reduce data variability. A one- to two-minute interval after speed stabilizes will suffice. The hull should be paddled in consistently deep water to avoid variable shallow water effects. Alternatively, you can paddle in water having any uniform depth, avoiding variable effects over the test range. Absent that, you can choose a test range with any bottom bathymetry as long as you paddle it in both directions: once with the full paddler complement, the second time with one paddler offline. This averages out any shallow water effects from the variable depth.

Taking a paddler offline in a C4's 2nd or 3rd seat likely entails a comparable speed impact. Comparing seats 1 and 2 is a bit more problematic since each seat provides different water access. In that case, you can conduct tests where paddlers swap their seats. For example, take whoever is in seat number 2 offline, swapping paddlers through that seat for each test. This removes any inherent biomechanical bias.

The essential test parameter to account for is paddlers' inclination to either speed up or slow down when one of them goes offline. Aside from accounting for human nature, maintaining uniform applied power from paddlers can be challenging. You can ask each paddler to maintain the same RPE (Rating of Perceived Effort) while underway. Or you can ask them to maintain a constant heart rate as presented on a heart rate monitor (HRM) during the test. Testing requires a thorough warmup to preclude heart rate drift during the tests. Heart rate is a more quantitative control than RPE. Plus, you can review heart rate data post-test to ensure the paddlers met this constraint. As you can see, there are many variables to manage to ensure a repeatable and fair test.

Finally, the SAR test protocol doesn't apply to tandems. For starters, it's challenging to paddle bow by yourself and maintain a heading.

Impact of Current

In our analysis we did not account for environmental factors such as current and wind. For C4 and OC-6 paddlers, you may find a location or time to conduct the SAR test with little or no wind. But not all test ranges are free from current. This can a problematic constraint for an OC-6 team. However, there is a way to overcome this problem.

If the test range's current doesn't change over the test time window, you can still measure SAR. As before, we'll consider the C4 case; the results apply to C6 hulls as well. Our hypothetical C4 is paddled left-to-right in the presence of a constant current having speed V_0, as depicted in Fig. 14.2. When the C4 travels with the current, its speed over ground increases to $V_0 + V$. V is the C4's speed absent any current. When the C4 travels against the current, its speed over ground decreases to $V_0 - V$. We'll denote speed over ground as $V_{measured}$.

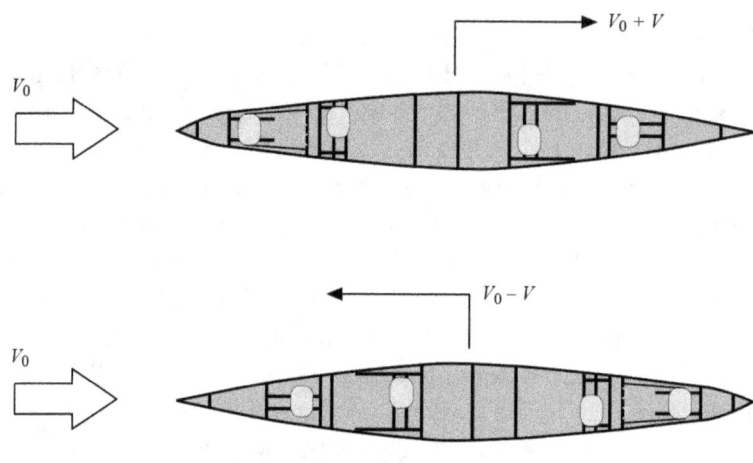

Figure 14.2: C4, with current.

To compute the SAR ratio, we require the hull speed attributable to the paddlers, not the measured speed over ground. The speed over ground includes "current power." However, we can measure speed over ground both with and against the current with a GPS. This provides two equations in two unknowns, V and V_0:

$$V_{measured,\,1} = V_0 + V \quad , \tag{14.9}$$
$$V_{measured,\,2} = V_0 - V$$

where the subscripts '1' and '2' denote the with and against current measurements, respectively. These are called *simultaneous equations*, which we can solve for both speeds by using a little algebra:

$$V = \frac{V_{measured,\,1} - V_{measured,\,2}}{2} \quad . \tag{14.10}$$
$$V_0 = \frac{V_{measured,\,1} + V_{measured,\,2}}{2}$$

Equations (14.10) yield the current-free hull speed V for the 4- and 3-paddler cases. The analysis then proceeds as above. The current speed V_0 pops out as a bonus.

But what if the current isn't directly head on or astern? This isn't a problem. Let's assume that the actual current is running abeam to the hull, with speed V_a as depicted in Fig. 14.3.

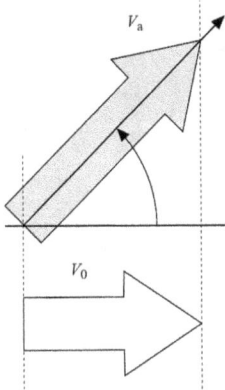

Figure 14.3: Current components.

In Fig. 14.3's plan view, the actual current is running at an arbitrary angle to our hull. But we don't care. The component of current along our C4's heading is still V_0. The speed V_0 is just the cosine of the angle between the actual current's direction and the boat heading, multiplied by the actual current speed V_a. Current has a direction and is thus a vector quantity. We're just using its vector component in the direction our hull is moving. This assumes that the current affects the hull equally when running abeam to the bow and abeam to the stern. We also assume that the C4's heading is the same paddled in both directions. This ensures the hull experiences the same current component. There are many things to account for if we wish to use the SAR ratio accurately and fairly. But despite all these qualifiers, it can still work.

Take-Aways

- Subject to the qualifiers and test controls outlined above, we can quantify a paddler's performance in an OC-6 outrigger or C-4 compared to their teammates with a metric based on paddling power.

- We developed a paddling Speed Above Replacement ("SAR") ratio you can use when selecting outrigger teams, or determining which seat is best for a C-4 paddler.

- We can compute the SAR ratio using GPS speed data from specific tests, as detailed above.

- Above a certain number of paddlers, the approach breaks down because the SAR ratio asymptotically approaches one. Changes in speed from taking one paddler offline will fall below our sensors' finite measurement precision.

CHAPTER 15

Leveling the Field

Introduction

One of the joys of paddlesport is seeing so many different people and boats at a race. Young and old; men and women; sleek hulls and less sleek hulls. We each have our paddling preferences. When given a chance, many of us will race solo kayak at one event, mixed tandem canoe (one man, one woman) at another, and so forth. The sport is very social, and it's always fun to mix things up. But we are, after all, racing. That competitive spirit bubbles up when we look at race results to evaluate our performance. We evaluate results to see how we compare our past performances and how we compare to other competitors.[78]

But how do we compare our race result versus competitors paddling different hull types, like a tandem versus a solo canoe? How do we compare paddlers of different ages? Or paddlers of different genders?

You can apply time adjustments to race results based on these differences; from now on I'll call these adjustments *gradings*. Gradings let us compare performances across different groups of paddlers on a more even footing. This chapter will consider how one racing organization implements a grading system to rank all paddlers. However, their time adjustment for women paddling solo canoes wasn't grounded in race data; there aren't that many race

---

[78] The word compete derives from the Latin *competere*, which means "to seek or strive for together." Competition is also about learning.

results for this group. We'll use physics to fill in the gap. The new grading helps level the field for all paddlers.

On Time Adjustments

Distance running embraced age- and gender-based race timing adjustments in 1989.[79] These gradings have been periodically updated since. The specific adjustments are recognized by USA Track and Field and World Masters Athletics. Similarly, the UK cycling Veterans Time Trials Association (VTTA) maintains "The Standards":

> [T]he Standard Tables are a set of target times for each distance which take into account a rider's age, gender, and type of machine. The aim is to allow all veterans to compete on a 'level playing field.' By comparing performances against their target time, rather than on actual time, the Standards provide fair competition for all ages and both men and women.[80]

While not the same as grading factors, the VTTA standards embody the same spirit.

The New England Canoe and Kayak Racing Association (NECKRA) has both down-river and flatwater Points Series where paddlers' race results are tallied over a season. As noted on their website *www.neckra.org*:

> "The NECKRA Points Series provides:
> • A fun way to track your racing
> • Encouragement to participate in more races
> • Competition on a more leveled playing ground
> Our... system has proven to be a good leveler of gender, age and boat type."

Their grading system adjusts finishing times based on paddler age and gender. They maintain adjustments for boat type as well. These include tandem and solo canoes, 3x27 pro boats, 16% recreational tandems, etc. Time adjustments are applied to each race's results.

A men's 3x27 tandem canoe, where both paddlers are under 50 years of age, receives no time adjustment in the NECKRA system. This is because an idealized men's pro boat performance is their reference for the canoeing time gradings. In a hypothetical race, the Point Series assumes that for canoe paddlers under 50,

• A pro boat (USCA 3x27 tandem) paddled by two men will finish in one hour,

• A mixed pro boat tandem (one man, one woman) in 1 hour 2 minutes,

• A women's pro boat tandem in 1 hour 4 minutes.

This is the structure of the NECKRA Point Series grading system. Some will paddle faster

---

79 https://github.com/AlanLyttonJones/Age-Grade-Tables ; http://www.legacy.usatf.org/Resources-for---/ Masters/LDR/Age-Grading.aspx

80 https://www.vtta.org.uk/content/0-standards

than this ideal, others more slowly. The stated goal of the NECKRA Points Series is to foster "Competition on a more leveled playing ground."

Let's consider an example of the NECKRA Points Series grading. For each paddler between 60 and 69 years of age, 1 minute *per hour* is subtracted from their adjusted finishing time. If a tandem men's canoe had one paddler aged 65 years and a finishing time of 1:30 (e.g., 1.5 hours), then

$$1.5 \text{ hours x 1 minute/hour} = 1.5 \text{ minutes}$$

If this tandem were paddling an NYMCRA Stock canoe, a further 3.5 minutes per hour (based on the original 1:30 finishing time) is subtracted from their adjusted finishing time,

$$1.5 \text{ hours x 3.5 minutes/hour} = 5.25 \text{ minutes}$$

The total adjusted time is then calculated as

$$1.5 \text{ hours} - 1.5 \text{ minutes} - 5.25 \text{ minutes} = 1:23:15 \text{ adjusted time}$$

Analogous time adjustments are applied to the finishing times of all racers who are NECKRA members. Similar computations are performed for all kayak finishers using kayak-specific gradings. Kayaks have a separate category in the NECKRA Points Series.

Once a race's adjusted finishing times have been computed, they are sorted. The fastest adjusted time is awarded 50 Series points for their performance, the second-fastest adjusted time receives 49 points, etc. The first boat to cross the finish line doesn't necessarily score 50 points. For example, a men's tandem may have had a mixed tandem (one man, one woman) on their stern in the finishing sprint. Everything else being equal, the time adjustment would award more Points to the mixed hull.

I served as the "Points Person" for NECKRA from 2008 through 2012, computing the Series results for their flatwater canoe and flatwater kayak races. The grading system proved to be a good leveler of gender, age, and boat type.

But there were a few exceptions. I was asked to consider whether their solo canoe ('C-1') gradings were reasonable. At the time, the men's C-1 grading was -3 minutes per hour, while the women's C-1 grading was -8 minutes per hour. These gradings were based on experience from past races, having been set by NECKRA's Competition Committee at their annual end-of-season meeting several years before.

I did a statistical analysis of race results that showed the men's C-1 grading should be adjusted to -4.5 minutes per hour. This grading has been in use since the 2010 season.

But what about the women's C-1 grading? Was it fair? There were so few women's C-1 finishers in past seasons that there was no way to draw a statistically meaningful conclusion either way.

Without sufficient timing data, we can use physics to level the field and determine a fair and supportable grading for women C-1 paddlers. To do this we'll leverage the tandem and solo canoe gradings cited above and a model developed in Chapter 1 that relates paddler power to steady-state hull speed.

Gradings and Paddling Power

From Chapter 1, we learned that a tandem canoe of length L with two identical paddlers that each supplies paddling power P produces a steady-state speed V_2 in quiet water (e.g., no current) via

$$2P = C_2 V_2^3 L^2 . \tag{15.1}$$

We've introduced a constant C_2 that synopsizes the tandem hull's hydrodynamics, thus greatly simplifying the analysis. If the two paddlers are a mixed team, we can denote the power supplied by the man as P_m and the power supplied by the woman as P_w. This allows us to express the mixed tandem's racing speed $V_{2,mix}$ as

$$V_{2,mix} = C_2' \sqrt[3]{P_m + P_w} L^{-\frac{2}{3}} \text{ for } C_2' \equiv C_2^{-\frac{1}{3}} . \tag{15.2}$$

Similarly, for a men's tandem

$$V_{2,men} = C_2' \sqrt[3]{2P_m} L^{-\frac{2}{3}} . \tag{15.3}$$

Now in high school algebra[81] we learned that "Distance equals velocity times time." Since all hulls in a hypothetical race will be traveling over the same course, the distance traveled by these canoes will all be the same. All that differs are the finishing times. Consequently, since things equal to the same thing are equal to each other,

$$V_{2,men} T_{2,men} = V_{2,mix} T_{2,mix} , \tag{15.4}$$

where $T_{2,men}$ and $T_{2,mix}$ denote the men's and mixed finishing times, respectively.

Recall that the NECKRA Points Series specifies a time adjustment for a hypothetical 1-hour race. For that 1-hour race, the time adjustment for a men's tandem is zero; for a mixed tandem, it is -2 minutes. This implies that in the absence of a time adjustment, the mixed tandem will finish in 1:02, or 1.0333 hours, or 1.0333 times the finishing time of the hypothetical men's tandem. Thus

$$V_{2,men} = 1.0333 \cdot T_{2,mix} . \tag{15.5}$$

Substituting the expressions above for the men's and mixed steady-state racing speed allows us to relate the men's and women's paddling power P_m and P_w inherent in these time adjustments,

---

81 We're using speed here rather than velocity; velocity is a vector quantity, which includes direction is as well. Apologies for the conflation.

$$\sqrt[3]{2P_m} = 1.0333 \cdot \sqrt[3]{P_m + P_w} \,. \tag{15.6}$$

If we assume that the men's paddling power P_m is one – this works because we're only interested in their relative paddling powers, not their absolute values – then

$$P_w = 0.8126 \cdot P_m \,. \tag{15.7}$$

This shows that the NECKRA mixed C-2 grading implies women exert about 81% of the power men do when paddling.[82]

We can check this result for internal consistency using NECKRA's women's tandem grading. As above, the steady-state speed for a women's tandem can be expressed as

$$V_{2,women} = C_2' \sqrt[3]{2P_w} L^{-\frac{2}{3}} \text{ for } C_2' \equiv C_2^{-\frac{1}{3}} \,. \tag{15.8}$$

For the hypothetical one-hour race, the men's tandem time adjustment is zero. For a women's tandem, it is -4 minutes, which means that in the absence of a time adjustment, the tandem will finish in 1:04, or 1.0666 hours, or 1.0666 times the finishing time of the hypothetical men's tandem. Thus

$$\sqrt[3]{2P_m} = 1.0666 \cdot \sqrt[3]{2P_w} \,. \tag{15.9}$$

If we once again assume that the men's paddling power P_m is one, then

$$P_w = 0.824 \cdot P_m \,. \tag{15.10}$$

This shows that the NECKRA women's C-2 grading implies women exert about 82% of the power men do when paddling. The difference between the paddling powers derived using the two different gradings – 0.8126 and 0.824, respectively – is about 1.4%. That's remarkably consistent. Check!

Leveling the Field

We now have enough information to level the field for all paddlers. We'll take a similar approach to our analysis above, but now use the NECKRA men's C-1 grading to derive a women's C-1 grading using the power factor we derived. Proceeding as before, but now with a men's solo canoe rather than a tandem,

$$V_{1,men} = C_1' \sqrt[3]{P_m} L^{-\frac{2}{3}} \text{ for } C_1' \equiv C_1^{-\frac{1}{3}} \,, \tag{15.11}$$

---

82 Note that this does not mean women exert 81% of the paddling power men do when paddling. To make any such determination, we would have to recruit a statistically meaningful sample of men and women paddlers, put them on an ergometer, and do testing. Our result means that the NECKRA Points Series grading, using the physics model outlined above, yields the 81% figure. When reading any technical article, always keep the underlying assumptions in mind when interpreting any result!

where V_1 and C_1 are the speed and hydrodynamics constant for a C-1, respectively. We have assumed that the tandem and solo hulls are the same length L. If not, any difference in length can be accounted for in the analysis.

Similarly, for a women's C-1,

$$V_{1,women} = C_1{}' \sqrt[3]{P_w} L^{-\frac{2}{3}}. \qquad (15.12)$$

Again, we note that "Distance equals velocity times time." Since all hulls in our hypothetical race travel over the same course, the distance traveled by the men's C-1 and women's C-1 will be the same. All that differs are the finishing times. Consequently, since things equal to the same thing are equal to each other,

$$V_{1,men} \cdot T_{1,men} = V_{1,women} \cdot T_{1,women}. \qquad (15.13)$$

Using our expressions for the Men's and Women's C-1 speeds in terms of their respective paddler's powers, this becomes

$$\sqrt[3]{P_m} \cdot T_{1,men} = \sqrt[3]{P_w} \cdot T_{1,women}. \qquad (15.14)$$

Using the value we derived for men's and women's relative paddling powers, $P_w = 0.824\, P_m$,

$$\sqrt[3]{P_m} \cdot T_{1,men} = \sqrt[3]{0.824 \cdot P_m} \cdot T_{1,women}. \qquad (15.15)$$

We're interested in determining the relative finishing times of the Men's and Women's C-1s; we already determined the relative paddling powers. Using the NECKRA Men's C-1 grading of -4.5 minutes per hour, the finishing time for a men's C-1 in a hypothetical race where a men's tandem finishes in 1 hour is 1:04.5, or 1.07666 hours. Solving the equation above for the Women's C-1 finishing time yields

$$T_{1,women} = \frac{T_{1,men}}{\sqrt[3]{0.824}} = 1.0666 \times 1.0766 = 1.148\,hr = 1:08.9. \qquad (15.16)$$

This suggests that the NECKRA Flatwater Points Series women's solo canoe grading should be -8.9 minutes per hour. Since the current grading is -8 minutes per hour, the two gradings differ by approximately 10%.[83]

NECKRA established the -8 minutes per hour time adjustment for women's C-1 paddlers some time ago because this number seemed reasonable. Our analysis shows that it is. Naturally, "reasonable" is based on an analysis that relies on other NECKRA gradings, which also seem reasonable. We've merely used science to show that the Flatwater Points

---

83 Note that this women's C-1 grading is derived using the men's C-1 grading. If the men's grading is updated for any reason, the women's should be also to maintain internal consistency.

Series is self-consistent, which is all you can ask for. And that physics can extend established systems to make up for the lack of a statistically significant ensemble of race results.

Take-Aways

- Using tools developed in this chapter, you can now grade paddling race finishing times using appropriate time adjustments to level the field for all paddlers.
- If there are not enough race results to calculate the grading for a particular paddler category you can use simple physics to derive it.
- If the physics model is based upon existing gradings, should those gradings be updated then the model should be updated as well.

Further Reading

https://github.com/AlanLyttonJones/Age-Grade-Tables (linked 10.15.2021).

http://www.legacy.usatf.org/Resources-for---/Masters/LDR/Age-Grading.aspx (linked 10.20.2021).

https://www.vtta.org.uk/content/0-standards (linked 10.20.2021).

Shawn Burke, "Where do Point Series time bonuses come from?", *The Competitive Paddler* (Spring 2011).

Part III

Movement

CHAPTER 16

What Fuels You

Introduction

I've wondered why canoes do what they do for some time. That curiosity led me to write all the previous chapters.

From my experiences training for downriver and flatwater races, I'm also curious why my body does what it does. Why should I train in a particular way? The common wisdom is that we need to perform aerobic overdistance paddling, intervals, and possibly tempo work to prepare for the racing season. But why? Why not just go out for a pint and be done with it? What goes on inside my body when I train or race? And does this relate in any way to how I *should* train?

It all comes down to how your body processes fuel and creates energy for movement.

Your working muscles are fueled by three different metabolic energy systems, sometimes called *pathways*: the *aerobic system*, the *anaerobic* or *lactate system*, and the *phosphate system*. Each plays a role corresponding to the intensity and duration of your paddling. The efficiency of these systems helps determine your endurance capacity.

The following overview of these controlling metabolic systems provides background and context for the curious paddler who wonders how and why the body performs and adapts to training. It also provides the foundation for the rest of the chapters in this book. And don't worry. You needn't be an exercise physiologist or biologist to read on.

The Aerobic System

The aerobic system, also called the oxygen system, utilizes fats (in particular, triglycerides) and sugar (in particular, glycogen) to fuel your muscles. It does so by producing a compound called *adenosine triphosphate* (ATP). ATP fuels muscle contractions, as we'll see below. The aerobic system is idealized in Fig. 16.1.

While this figure may look like a biochemistry gobbledygook, it represents much of what happens in our muscles when we paddle. We noted that there are two raw fuel sources for the aerobic system: fats and sugars. Your body stores dietary fats as triglycerides. These are processed during exercise in several intermediate steps. In the first step, triglycerides are combined with water (H_2O) to produce fatty acids and glycerol, as shown on the left-hand side of Fig. 16.1. The fatty acids are further refined in the muscles via a process called *beta oxidation*.

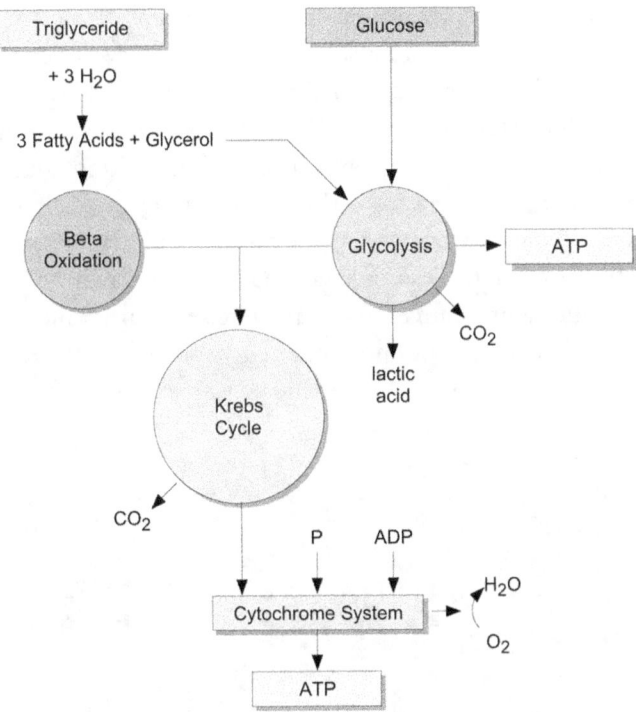

Figure 16.1: Combined energy pathways.

Glycogen, a long-chain sugar molecule, is stored in the liver and muscles.[84] During exercise, glycogen is processed in the muscles via *glycolysis* to produce ATP. The muscles may use this ATP; glycolysis is a fast reaction compared to beta oxidation. The glycerol released during fatty acid production is also processed to produce ATP via glycolysis, as shown in the figure.

Both beta oxidation and glycolysis produce intermediate compounds that are necessary

---

84 This happens whether you are carbo loading before a race or not.

for muscle metabolism. These compounds enter structures within the muscles' cells called *mitochondria*. Mitochondria are the engines that produce ATP at the cellular level through the *Krebs Cycle* and the *Electron Transport* or *Cytpchrome System*. The cytochrome system combines the byproduct of muscle metabolism, *adenosine diphosphate* (ADP), with phosphorous (P) to produce ATP. As Fig. 16.1 indicates, the aerobic system is where oxygen (O_2) is "burned," combining with free hydrogen ions to produce water, as shown in the lower right-hand corner of the diagram.

The muscles continuously process ATP to release energy for muscle contractions. This energy is produced when a phosphorous atom is released from ATP (the "tri"-phosphate), producing ADP (the "di"-phosphate). The precursor compound ADP is always present in the muscles, waiting to be recycled in the cytochrome system to produce ATP. All that is required is fuel, either as fats or carbohydrates. These are processed through the linked pathways of beta oxidation, glycolysis, and the combining Krebs cycle.

As with any oxidative process, there are byproducts, much like a car produces emissions when it burns fuel. The byproducts of muscle respiration – *carbon dioxide* (CO_2) and *water* (H_2O) – are removed by the lungs and carried through the bloodstream, respectively.

The body's carbohydrate store is limited. The store of fats that can feed beta oxidation is comparatively unlimited no matter how lean you are. The "fat" and "glucose" systems work simultaneously. But their contributions to the energy supply are different and depend on your level of exertion and how you train. Fats provide the dominant energy source primarily during low-intensity exercise. For example, as you sit and read this chapter, your body is mostly burning fat as its energy source. As your exercise intensity increases, carbohydrate oxidation via glycolysis becomes the dominant energy source, as idealized in Fig. 16.2.

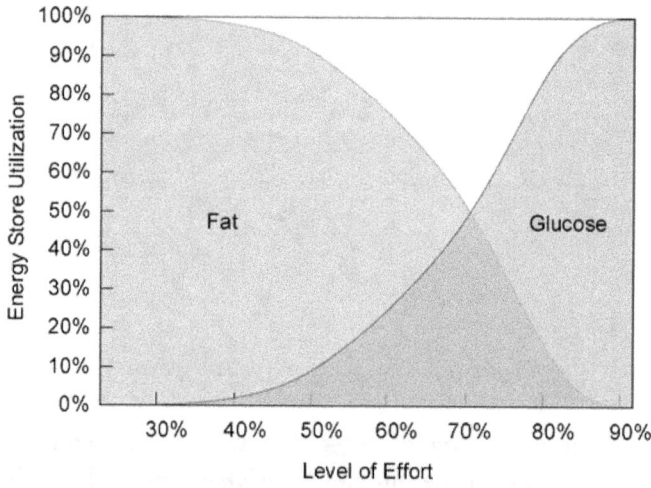

Figure 16.2: Aerobic fat and glucose utilization during steady exercise.

The body can store enough carbohydrates in the muscles and liver to provide approximately 90 minutes of glycolytic energy alone. When your carbohydrate stores are depleted, fat burning increases to compensate, and the ability to sustain high levels of effort decreases. Fat oxidation always requires a small amount of carbohydrates to keep the fires lit, which is why a marathoner hits the wall around the 90-minute mark unless they replenish their glycogen stores during the race. When the body's carbohydrate stores are depleted during long races or workouts, we can prolong glycolysis by drinking carbohydrate drinks or eating modest amounts of easily digested high-carbohydrate foods.[85]

Through exercise, you can train your muscles' metabolic systems to use energy more economically. The number and density of muscle mitochondria increase with appropriate training, providing more engines to burn fat and glucose. The metabolic pathways become more efficient as well. A well-trained athlete can oxidize fat for a longer time than one with less training. Training shifts the curves in Fig. 16.2 to the right and thus saves (some) glycogen for high-intensity efforts like jumping wakes and finishing sprints.

In addition to metabolic adaptations, aerobic exercise increases the number and density of the tiny blood vessels within the muscles called capillaries. This change is visible under a microscope after the first few weeks of aerobic training – just don't try this at home! As a result, the heart will supply more oxygen-rich blood to the muscles and more efficiently remove the chemical by-products of exercise, including carbon dioxide, water, lactic acid, and free hydrogen ions.

A long-term program of aerobic exercise also increases the heart's strength and stroke volume. The well-trained heart can pump more blood with each beat, with a corresponding lower heart rate for the same level of effort and a lower resting heart rate.

A few other tidbits about the aerobic system that may shape your training:

- As you begin to exercise, it takes at least 2 to 3 minutes before the heart, lungs, and circulatory system are fully functioning, and the aerobic system is fully engaged. This is one reason why you need to warm up before training or racing.

- After prolonged, strenuous exercise, it takes as long as 48 hours to replenish glycogen stores in the liver. *Muscle* glycogen, however, can be replenished in a few hours. Consequently, long overdistance workouts –– approximately 2 hours or longer – should be followed by a recovery day of light exercise or a strength-training day. That said, many ultra-distance racers perform "sandwich" aerobic workouts on consecutive days, say three hours one day and five hours the next. This acclimates the body to partial recovery in preparation for long races like the AuSable Marathon or Yukon River Quest.

- Aerobic energy is synonymous with oxygen uptake capacity and mitochondria density. The body adapts to training stress over time with increased number and density of aerobic muscle mitochondria from a baseline of 3.5% muscle mass to as much as 5% muscle mass. Extensive aerobic training can increase aerobic endurance capacity by approximately 50%. This adaptation takes about four years to plateau. Sorry, but great endurance athletes aren't created overnight!

- Increases in mitochondria number and density and capillarization occur *in the muscles that we stress* in aerobic exercise. So, to develop aerobic capacity for paddling, you should paddle. This is called *training specificity*. Running and cycling, for example, won't by

---

85 There are other approaches to training and racing based on low-carb diets, but they are outside the scope of this chapter.

themselves maximize paddling aerobic fitness, but they certainly increase cardiopulmonary efficiency. You are what you eat; you can best do what you train for.

The Anaerobic System

Glycolysis produces lactic acid, a metabolic byproduct we conveniently ignored when discussing Fig. 16.1. The liver reprocesses lactic acid via the Cori Cycle, which produces glucose to fuel the muscles. The Cori Cycle is depicted conceptually in Fig. 16.3. Glycolysis and the Cori cycle do not require oxygen to produce ATP; the Cori Cycle is *anaerobic*, i.e., it functions without oxygen. Note that glycolysis and the Cori cycle produce ATP more quickly than the aerobic system since the anaerobic pathway entails only about a dozen chemical reactions.

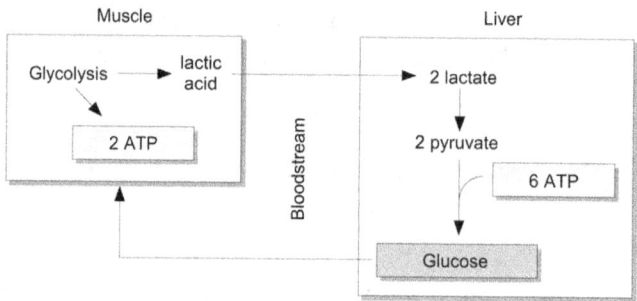

Figure 16.3: The Cori Cycle.

At the start of exercise, the energy supply is anaerobic, irrespective of intensity. We are fueled first by the phosphate system (which we'll discuss later) and then by the anaerobic system. For exercise lasting around 1 to 3 minutes, such as $400m$ and $800m$ runs, the energy supply is predominantly anaerobic; recall that it takes 2 to 3 minutes before the aerobic system "warms up" entirely and provides ATP. The anaerobic system also supplies the energy when we increase pace during surges and finishing sprints. During steady-state exercise, the Cori Cycle continues operating in the background, even at very low levels of exercise intensity. The Cori Cycle can recycle lactic acid completely *below a blood lactate concentration of around 4.0 millimoles per liter*. At and below this concentration, lactate is providing fuel for your working muscles.

Beyond a certain level of exertion, this recycling process cannot keep up with the muscles' lactic acid production. Recycling hits a threshold at a blood lactate concentration of approximately 4.0 millimoles per liter. As lactate increases to around 7.0 millimoles per liter, beta oxidation (fat metabolism) essentially stops, and the muscles are increasingly fueled by recycling lactate into glucose via the Cori Cycle. At extremely high exercise intensities, our blood lactate concentration can peak at approximately 25 millimoles per liter.

Lactic acid production ultimately limits anaerobic power. Lactate accumulation increases muscle acidity. This state of muscle acidity is called *acidosis*. Acidosis damages the muscle cell walls, causing leakage through the cell membrane into the blood. Acidosis also interferes with and damages the system of aerobic enzymes within the cells, decreasing aerobic endurance capacity. Adding insult to injury, muscle contractions at very high lactate concentrations become more difficult because of a lack of ATP. Anyone who has run an all-out 400*m* sprint will be familiar with this state. It feels like a bear jumped on your back at the 300*m* mark.

With increasing acidosis, an enzyme produced by glycolysis blocks lactic acid production. If not for this enzyme, further increases in muscle acidity would damage muscle protein and mitochondria. The body does not learn to "tolerate" high lactic acid production; lactic acid production is self-limiting. The body can adapt to intense workouts and racing with training and learn to operate within tolerable lactate levels. The level of exercise intensity where lactic acid recycling matches its production is called the *lactate threshold*. The lactate threshold ultimately determines race performance. We'll review how to measure this threshold for paddling in Chapter 18.

After strenuous anaerobic exercise, it can take days before the body sufficiently recovers and regains maximum aerobic capacity. When exercise is performed repeatedly at too high an intensity – without sufficient recovery between bouts of hard anaerobic exercise – aerobic and anaerobic endurance capacity decrease considerably. This often leads to problems collectively known as *overtraining*. Recovery time – 24 hours or more – helps muscle cells normalize after hard anaerobic workouts and races. Recovery is a valuable tool when developing and undertaking training programs that include hard intervals, high intensity "threshold" exercise or tempo training, and racing.

A few other tidbits about the anaerobic system that may shape your exercise:

- It takes the body about 25 minutes to remove half the accumulated blood lactate from maximum exertion when resting. 95% of lactate is removed after about 1 hour and 15 minutes of rest.

- Lactate is removed from the blood and muscles much more quickly when you perform light, continuous exercise at the end of a workout, rather than just stopping. This *active recovery* is ideal for your cool-down following a race or hard workout. A cool-down of light paddling will remove 90% of blood lactate in approximately 20 minutes. Yeah, that's a long cool down. But five or ten minutes should be possible.

- It is commonly assumed that lactic acid is responsible for increased recovery times after intense workouts. However, free hydrogen ions and other byproducts of anaerobic exercise also delay recovery.

The Phosphate System

The phosphate system supplies energy directly to the muscles. It is fed by the aerobic and anaerobic pathways as described above. The phosphate system also drives fast and powerful muscle contractions yet does not use oxygen. This is fortunate since very intense muscle contractions squeeze off the oxygen-rich capillary blood supply. The phosphate system does not use glucose or produce lactic acid; it supplies anaerobic *alactic* energy. It is the fastest muscle energy pathway because it requires the fewest chemical reactions.

Muscle contractions are controlled locally by our friend ATP, which breaks down in the muscles to produce ADP and energy:

$$ATP \rightarrow ADP + Energy \, . \qquad (16.1)$$

Before we exercise, the muscles themselves have a store of ATP that can be called upon for short bursts of maximum effort – up to approximately 3 seconds. The phosphate system then kicks in and provides energy for maximum effort for another 10 seconds. This is the energy source a sprinter calls upon in a $100m$ dash. The phosphate system is vital during short-duration sprints, race starts (hey, free energy!), and strength training.[86]

A compound called creatine phosphate (CP) provides the necessary phosphorous (P) to drive the phosphate system, re-synthesizing ATP in the muscles once their stores are exhausted. The phosphate system replenishes 70% of ATP and CP within 30 seconds; these compounds are completely replenished in 3 to 5 minutes. The phosphate system is trained by hard, short efforts, alternated with periods of complete rest. Think weight training when you're lifting heavy, which relies solely on the phosphate system, or sprint training. Rest periods should be long enough for ATP and CP to re-synthesize. Note that high lactic acid concentrations delay CP replenishment. Consequently, we shouldn't combine anaerobic training sessions with pure sprint speed workouts.

After several months of sprint training the muscles' ATP and CP stores increase by 25% to 50%. However, the ultimate speed we can achieve when sprinting isn't limited by the muscles' ATP store or the efficiency of its re-synthesis. Speed is a function of technique, which is determined by the central nervous system and our neuro-muscular pathways.[87]

Muscle Metabolism, Performance, and Training

The body employs all three energy systems described above during exercise. Yes, all of them. Their *proportional engagement* depends on the length and intensity of effort. The relationship between exercise duration and the fractional contribution from the various energy systems is conceptualized in Fig. 16.4. Each bar on the plot represents a maximum effort for the corresponding time interval on the bottom axis. As indicated by the figure, short sprints are anaerobic and alactic, primarily engaging the phosphate system. Assuming you've warmed up thoroughly, a starting sprint minimally engages the aerobic system. However, the phosphate and anaerobic systems dominate the energy supply. For longer races, the aerobic system becomes the dominant source of energy. The changing contributions from the energy systems vary continuously. It is not an abrupt transition from one system to the next.

---

86 Some authors conflate anaerobic exercise with anaerobic alactic exercise. This leads them to assert that endurance athletes shouldn't do strength training because it's anaerobic, and anaerobic exercise produces lactic acid. Strength training is alactic, and therefore doesn't produce lactic acid.

87 Speed also depends on our body's ratio and distribution of fast- and slow-twitch muscle fiber. Successful sprint specialists have a preponderance of fast-twitch muscles. Grinders have more slow-twitch fibers.

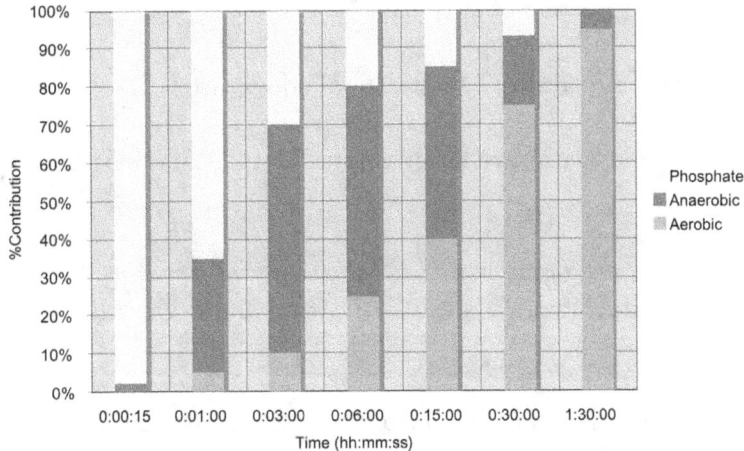

Figure 16.4: Energy utilization vs. time (after Janssen).

This graph suggests that you should train each energy system in proportion to its engagement for a goal race's length. In other words, a paddler training for a long race should spend most of their training volume developing their aerobic system, especially to improve the efficiency of fat oxidation, and far less (but not zero) training their anaerobic system. The phosphate system is much less relevant to train in isolation for these paddlers – except for starts. As the season progresses and we transition to shorter races, training can incorporate more anaerobic and sprint work dictated by new goal race distances. Racers training all along for shorter events will already be doing this. And Fig. 16.4 shows why.

Take-Aways

- Your working muscles are fueled by three different metabolic energy systems: the *aerobic system*, the *anaerobic* or *lactate system*, and the *phosphate system*.
- Each metabolic system plays a role corresponding to the intensity and duration of paddling. The phosphate system, which is always engaged, plays the starring role in race starts and when lifting weights. The aerobic system burns sugars in the presence of oxygen to power working muscles. The anaerobic system reprocesses blood lactate to power working muscles at high exercise intensities.
- The efficiency of these systems helps determine your endurance capacity. Consequently, you can balance training these systems in line with your performance goals, and the duration and expected intensity of races in your paddling season.

Further Reading

Anderson, Owen, "Periodization Training Technique: If you want to improve your performance, you can't train the same way all the time," http://www.pponline.co.uk/encyc/0147.htm.

Bishop, D., Bonetti, D., & Dawson, B., "The effect of three different warm-up intensities on kayak ergometer performance." *Medicine and Science in Sports and Exercise*, **33**, pp. 1026-1032.

Croston, Glenn, "Beta oxidation of fatty acids,"

http://www.biocarta.com/pathfiles/betaoxidationPathway.asp.

Croston, Glenn, "Feeder pathways for glycolysis," http://www.biocarta.com/pathfiles/feederPathway.asp.

Endicott, William T., *The Barton Mold, A Study in Sprint Kayaking*, http://www.davey-hearn.com/Coaching/Technique/The%20Endicott%20Files/The%20Barton%2 0Mold/the_barton_mold.htm.

Higdon, Hal, "Ultramarathon training," http://www.halhigdon.com/ultramarathon/ultra-marathon2000.htm, 2000.

Isaka, T., & Takahashi, K. "Effects of off- and pre-season training on aerobic and anaerobic power of kayak paddlers," *Medicine and Science in Sports and Exercise*, **29**(5), Supplement abstract 1242, 1997.

Janssen, Peter, *Lactate Threshold Training*, Human Kinetics, Champaign, IL, 2001.

Maffetone, Philip, *Training for Endurance*, David Barimore Productions, Stamford NY, 1996.

Noakes, Tim, *Lore of Running*, 4th Edition, Human Kinetics, Champaign, IL, 2003.

Nolte, Volker, *Rowing Faster*, Human Kinetics, Champaign, IL, 2005.

Pfitzinger, Pete, "How to speed up recovery from racing," http://www.copacabanarunners.net/i-recovery-racing.html, 2004.

Pfitzinger, Pete, "Finding your optimal training / recover ratio," http://www.copacabanarunners.net/i-training-recovery.html, 2004.

Schulman, Deborah, "Fuel on fat for the long run," http://www.marathonguide.com/training/articles/MandBFuelOnFat.cfm, 2000.

Sleamaker, Rob, and Browning, Ray, *SERIOUS Training for Endurance Athletes*, 2nd Edition, Human Kinetics, Champaign, IL, 1996.

Szanto, Csabo, "Daily training program for advanced athletes," http://www.canoeicf.com/default.asp?Page=1605&MenuID=Development%2F1012%2F0.

http://www.nismat.org/physcor/max_o2.html, "NISMAT Exercise Physiology Corner: Maximum Oxygen Consumption Primer," 2005.

http://www.bbc.co.uk/scotland/education/bitesize/higher/biology/cell_biology/respiration2_rev.shtml, "Stages of aerobic respiration."

http://biocarta.com/pathfiles/h_etcPathway.asp, "Electron transport reaction in mitochondria."

CHAPTER 17

Power to the Paddlers

Introduction

Some of us supplement our paddling with strength training; many of us do not. I've felt that strength training has helped me with starts and jumps when racing. But my assessment is purely qualitative; my stroke feels "stronger" if I've been lifting conscientiously, but what does stronger mean?

Many excellent paddlers pass me once we're past the start. How much has strength training benefited me downstream? And equally important, does the time I spend on strength training provide enough benefit to keep doing it? Am I doing it "right"? And for paddlers, what is right?

In this chapter, we'll consider a strength training method that aligns with the science of paddling. The underlying philosophy is grounded in both physics and exercise physiology, borne out by my own experience. Rather than training for strength or muscle mass, this approach focuses on developing power. To make a case for it, we'll review more conventional perspectives on strength training's functional benefits. Then, we'll add a pinch of physics and see what happens when we combine them.

On Strength Training

My introduction to strength training came in college. I knew lots of jocks, mostly football players. I would diligently follow them to the gym and muck about with free weights and machines. The gospel was "three sets of ten reps"[88] for each exercise. I gained a bit of muscle doing this. My scrawny upper body became more balanced with my lower body. And that was about it.

When I decided to get serious with off-season training more than a decade ago, I researched strength training for paddlers. I mostly found prescriptions for specific, functional exercises. Then, in Tudor Bompa's books *Periodization* and *Periodization Training for Sport*, I found detail – lots of detail. Bompa laid out the benefits of weightlifting performed at specific fractions of one's single repetition maximum[89] ("1RM") for exercises corresponding to the function(s) one wished to train. You'll find this functional breakdown illustrated in Fig. 17.1. Note that we're not yet focusing on any one exercise; that will come later. For the moment, let's generically consider any strength training exercise.

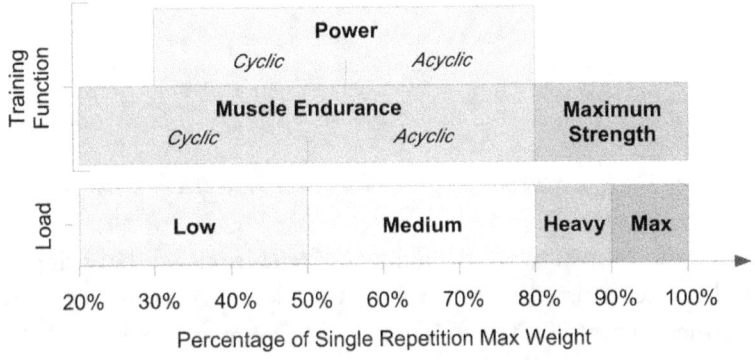

Figure 17.1: Functional benefits of weight training vs. percentage of 1RM (after Bompa).

This figure summarizes the functional benefits of performing an exercise at various fractions of 1RM. Lifting a load ranging from 80-100% of 1RM benefits our maximum strength for that exercise. From 30% to 80% of 1RM benefits "power"; Bompa divides power into acyclic movements and cyclic movements. Acyclic movements are performed singly or infrequently, like jumping or lunging during a game of basketball. Cyclic movements are frequent and periodic, like paddling.

I also learned about the relationship between percentages of 1RM and the number of repetitions I could perform in a single set before I couldn't do another. This is known as "reps to failure" for that exercise. A plot of reps to failure appears in Fig. 17.2.

In Fig. 17.2, we see that you can lift 100% of your single repetition maximum once. That should be obvious. You can lift 90% of your 1RM four times, 72% of your 1RM ten times, and so forth. And when I write four times, that means you can't perform that same exercise a fifth time within the same set. Ditto for ten repetitions at 72%; if you can do an eleventh repetition, you're lifting less than 72% of your 1RM.

---

88 For the uninitiated, "rep" is short for *repetition*, a single instance of an exercise movement such as lifting a weight. A "set" is a number of consecutive repetitions of one exercise.

89 Single Repetition Maximum is the maximum weight one can lift once and only once for a given exercise.

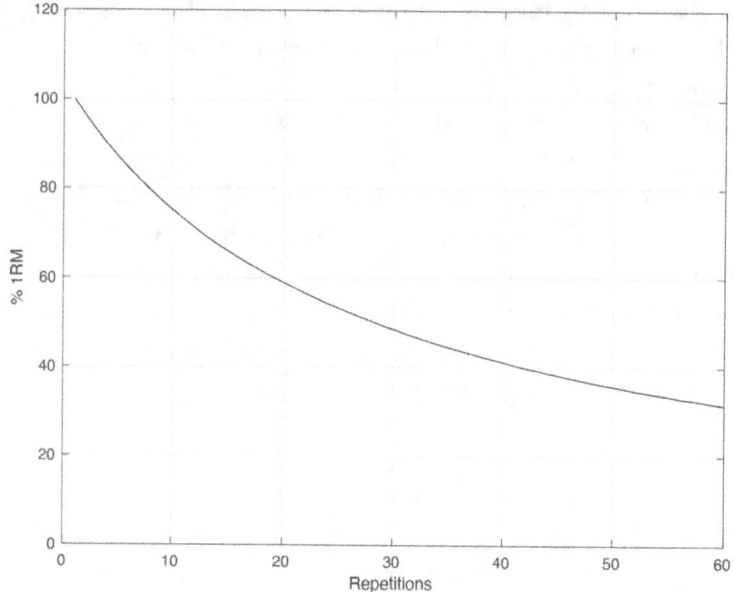

Figure 17.2: Repetitions to failure for loads that are a percentage of 1RM.

I found that the relationship between the number of repetitions I did in a single set to failure vs. weight lifted tracked this curve remarkably well down to about 50% of 1RM.[90] So I diligently performed lifts at 80-90% to increase maximum strength; 70-80% for "power," and so forth. I got stronger and a little bigger. But aside from "feeling stronger," it was unclear how this work benefited my performance on the water.

After undertaking a pre-season program like this for four years, whatever nebulous benefits I reaped from horsing around all that weight plateaued. My muscles didn't feel as tired late in races as they once did (though the rest of me did!) I started looking for something more impactful. A friend and fellow paddler pulled me aside and said, "I've got the workout for you."

The workout consisted of four (mostly) paddling-specific exercises, done using bodyweight only, performed as a circuit.[91] I performed the circuit three to five times per session. The exercises comprised pullups, ab wheel rollouts, inverted rows,[92] and pushups. Doing this workout 3-5 days per week helped, and the workout only took 20-25 minutes to boot. I think a good deal of the benefit comes from how these functional exercises emphasize core engagement.[93]

---

90 Below that, predicted performance didn't match my experience in the gym.

91 In a circuit, you perform one exercise set, immediately followed by sets of other exercises, and so on until you complete all exercises. You don't take a break between sets.

92 An inverted row is sometimes called a reverse pushup. It's like a bench row, but with no bench, and upside down to boot. Heels on the floor, body in a plank, back just above the floor, arms extended upwards to a bar (or suspension trainer). Hold the plank and pull, pivoting at the heels. Lower yourself down. Repeat.

93 "Traditional" strength training often emphasizes exercises that isolate muscle groups. Contrast a bench press with a pushup. While the bench press is a great powerlifting move, it primarily works the triceps,

I thought that was it. And yet, something was still niggling at me. I'm an engineer; I'm driven to ask, "Why?" I understood how the exercises were paddling specific. But aside from general feeling a further benefit on the water owing to the exercises' specificity, could I get even more bang for the buck if I did these exercises differently?

Paddling and Power

So how does strength relate to paddling speed? In Chapter 1, we learned that the steady-state cruising speed V of a hull is proportional to the cube root of the average power P that the paddlers put into the water:

$$V \propto \sqrt[3]{P} \times L^{-\frac{2}{3}}. \tag{17.1}$$

Note that Equation (17.1) does not indicate that speed is proportional to strength. Strength coincides with the force we apply through the paddle to the water. Since speed is not proportional to strength, this suggests that merely getting stronger and thus applying a lot of force to the water through a paddle is not sufficient to move a hull fast. Instead, we see that speed is proportional to power. This reflects the observation from Chapter 6 that propulsive impulse is maximized if peak paddle force occurs simultaneously with peak relative velocity between hull and paddle. So then, what is power?

Power P is the rate of doing work or the amount of energy transferred per unit time. Mathematically, power is equal to the product of an applied force F times the velocity v by which this force is applied:[94]

$$P = F \cdot v. \tag{17.2}$$

Since average velocity v is equal to distance D divided by elapsed time Δt, we can alternatively represent power as

$$P = \frac{F \cdot D}{\Delta t}. \tag{17.3}$$

Force times the distance over which the force is applied is defined as *work*. This is Equation (17.3)'s numerator, so power P is work per unit time.[95] Equation (17.3) also shows that for a given paddle force F and stroke distance D, paddling power P increases if the stroke duration Δt decreases. In other words, you put more power in the water for a given paddle force if

---

anterior deltoids, and pectoral muscles. The bench press secondarily activates adjacent muscles in the upper body. A pushup activates these muscles as well. But a pushup also engages the core, along with much of the biomechanical chain from heels to shoulders.

94 While power is a scalar quantity, both force and velocity are vectors. As we've noted before, this means that force and velocity have a direction. We'll assume that force and velocity are aligned; we'll only consider their scalar magnitudes. Or, if you prefer more mathematical rigor, note that the expression for power is a dot product. The dot product of two vectors yields a scalar. It's your choice.

95 And work has units of energy, like Joules or Calories. Very convenient.

you exert this force quickly. Which makes you go faster, per Equation (17.1), reflecting the importance of stroke cadence and duty cycle discussed in Chapter 7.

So "larger" power is the quick application of a force. When paddling, you need to do this over and over if you want to go faster. Equation (17.2) then begs the question of which element you train: force or velocity? What if you could train paddling power directly without having to deconstruct it into components?

One of my favorite tech blogs had a review for a new exercise gizmo that measures the velocity[96] of your movements during strength training. For a given exercise, you know the weight you're moving, you'll know the power exerted during that exercise. There are many velocity sensors on the market. Mine has a small Bluetooth 4.1-connected sensor that I could place atop a weight stack. You can also insert it in a wrist band for bar-based lifts or attach it to a body mount for pushups, pullups, inverted rows, and the like. These last two configurations capture bodily movement, and thus velocity at the attachment point.

The sensor sends average velocity data to a connected smartphone app. The app lets you select an exercise, enter the weight to be lifted, and then presents a "Start" button to initiate data recording. The app reports either the lift's velocity (in *m/s*) or the lift's power (in *Watts*) for every rep. You can observe results after each rep, along with a bar graph of rep velocity over a set.

OK, great; there's a gizmo. Engineers like gizmos. Now what?

Velocity Based Training

In researching velocity sensors, I learned about a branch of strength training called "Velocity Based Training" or "VBT." VBT's thesis is that specific training ranges based on weights and velocities produce specific functional benefits. These functional ranges reminded me of Bompa's prescription, which bases training ranges on percentages of 1RM lifts. But VBT embodies one key difference: it focuses on velocity rather than 1RM.

In conventional strength training, we determine our 1RM then lift at some fraction of 1RM until that weight increases. Perhaps we gain muscle mass – a phenomenon called hypertrophy – or achieve some other goal to increase it.

But on any given day, an athlete can have a different 1RM. Previous hard training, irregular sleep patterns, and stress can lower 1RM. Our single repetition maximum is not a fixed quantity. On a given day, prescribing three sets of ten reps at 70% max weight may result in the athlete lifting 80% of their 1RM. This won't achieve the desired training effect. Do this enough, and injury or overtraining can follow. Or prescribing lifts at 70% of max may correspond to 60% on a given day. This too won't produce the desired training effect.

Alternatively, we can train by moving a weight at a specific target velocity. We adjust the weight so that a maximal effort results in that desired velocity for a series of reps. Consequently, you will always be training "correctly" for a desired functional result irrespective of the weight lifted. This holds even when your 1RM changes up or down.

---

96 The sensor measures velocity components using a MEMS accelerometer and reports back speed in the direction of travel. Again, velocity is a vector, while speed is a scalar. But in this chapter, we'll occasionally conflate the two. Sorry!

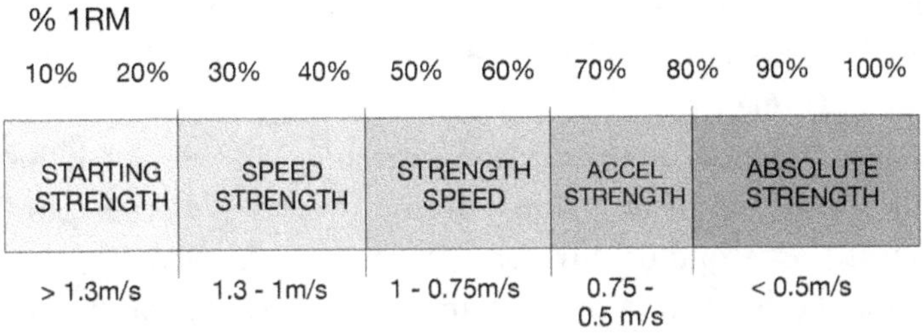

% 1RM

10%	20%	30%	40%	50%	60%	70%	80%	90%	100%

STARTING STRENGTH	SPEED STRENGTH	STRENGTH SPEED	ACCEL STRENGTH	ABSOLUTE STRENGTH
> 1.3m/s	1.3 - 1m/s	1 - 0.75m/s	0.75 - 0.5 m/s	< 0.5m/s

Figure 17.3: VBT velocity ranges and associated functional benefits.

There are now generally accepted velocity ranges that target specific functional benefits. These are shown in Fig. 17.3. In this figure slower velocity ranges are to the right, while faster velocity ranges are to the left. There are two particularly interesting velocity ranges: Speed-Strength, and Strength-Speed. Mann[97] defines Strength-Speed as "moving a moderate weight as fast as possible." Strength-Speed has a higher force development rate compared to accelerative strength. Mann further defines Speed-Strength as "moving a lighter weight as fast as possible." He notes that "[t]his trait has the second-highest rate of force development of all the traits." Racers especially know that we're quicky exerting a force through the water, so training somewhere in this velocity range is relevant for paddlers. I set out to try it.

Using the sensor, I performed the following exercises: Lat pulldown, inverted row, pushup, and overhead cable press. My goal was an average velocity of 0.75 *m/s* over a set, corresponding to the lower velocity end of the Strength-Speed range above. I chose this as a starting point since it led me to select weights that provided enough resistance to make the lifts "interesting." Faster velocities like 1.0 *m/s* just felt too easy given their lower weights.

For the overhead cable press and lat pulldown, I determined the weight I could move at 0.75 *m/s*. After that, I moved the weight as close to 0.75 *m/s* as possible for each lift while maintaining good form. No cheating! The sensor's smartphone app displays the average velocity for each rep. This instant feedback is both fascinating and encouraging; you know exactly where you're at for each lift across a set.

How many repetitions per set, and how many sets, should you perform? The answer involves a bit of art informed by science. From Chapter 16, recall that the phosphate system, which ultimately fuels the muscles, has a store of "free" energy in the form of Adenosine Triphosphate (ATP). We deplete the ATP stored in our muscles over the first 10-12 seconds of exercise, say performing a sprint or lifting weights. After that, the other metabolic systems take over. It takes about 2-3 minutes of rest to replenish this "free" energy store in the working muscles. I reasoned that I should perform a specific exercise for 10-12 seconds[98] at the target velocity for a given exercise and resistance, then move on to the next exercise (and its different working muscles). This ensures we deplete the working muscles' ATP store

---

97 Mann, Bryan, Developing *Explosive Athletes: Use of Velocity Based Training in Training Athletes, 3rd Edition*, Ultimate Athlete Concepts, Michigan USA (2016).

98 Yes, I used a timer. Yes, I'm an engineer.

over each set. A circuit lasts long enough that the muscles replenish ATP before starting the next set of a given exercise. I rested between circuits to ensure my working muscles were ready for the next one.

The 10-12 second set duration almost always resulted in 8 reps for the 0.75 *m/s* target velocity. I usually saw the rep velocity slowly decline over a set. I often found that my "best" (i.e., fastest) rep was the second or third one. I interpret the velocity decline to depletion of the ATP stores and other neuromuscular and biochemical factors coming into play that limits a muscle group's ability to continue doing work at that rate.

The decline in rep velocity is excellent feedback. If you're not hitting your target average velocity over a set, you can adjust the weight up or down to get the desired training benefit. After a few trial sessions, I established eight reps/set as my goal at 0.75 *m/s*. Again, these are not sets done to failure like in classic strength training. They are done within a defined time range for reasons noted above, with an eye to monitoring degradation of rep velocity over the set. VBT is about quality, not quantity. I do between three and five sets per session. I continue lifting as long as I can hit the target velocity in each set, with no loss in form.

As to the bodyweight exercises – pushups and inverted rows – I've been able to do sets at the average target velocity of 0.75 *m/s* for inverted rows and about 0.6 – 0.65 *m/s* for pushups. For those of you who have done plyometric pushups – such as pushups where you clap your hands at the top of every repetition – this velocity range has a similar feeling during the concentric muscle contraction. As I progressed with inverted rows, I added a weighted vest to increase resistance. You can do the same for pushups. Again, the goal is to achieve an average target velocity, not a particular weight. The weight you move will systematically increase over time. If you consistently exceed your average target velocity, it's time to increase the resistance.

Toward Maximizing Power

Over one month of VBT sessions, I saw modest but consistent, measurable gains in weight moved at 0.75 *m/s*. That was encouraging. So naturally I wondered if there is a way to determine experimentally whether 0.75 *m/s* is the correct average velocity that maximizes power. Remember, the cruising speed of a hull is proportional to power's cube root.

Recall that power is the product of weight, which is a force, times velocity. For light weights, I could easily exceed my 0.75 *m/s* velocity target. For heavier loads I slowed down to as little as 0.28 *m/s*. At this end, I was in the "Absolute Strength" velocity range of Fig. 17.3. Now it's nice to work at a target velocity prescribed in a book or technical paper. But is there a relationship between velocity and power, or weight lifted and power, and does this relationship indicate an optimum velocity to maximized power? In other words, if I varied both velocity and force, what combination yields the most power?

If you can move a light weight very fast, you might be generating less power than moving a somewhat heavier weight a little less fast. Also, slowly lifting a heavy weight might generate even less power than moving a light weight fast. To provide contexts for these scenarios, I constructed a *power curve* of my lat pulldowns.

After calibration, I performed sets of 8 reps, progressively adding 10 *lb.* between sets. I started at 120 *lb.* and progressed to 210 *lb.*, which is the limit of my machine. At each weight, and for each rep, I performed the exercise as fast as possible while maintaining good form. I did no other exercises during this session. Between each set, I rested for 3 minutes to ensure ATP replenishment. I limited myself to 4-6 repetitions at the highest weights since I could see my velocity start to drop off with each rep.[99] While tiring, this workout was fun because it generated data.

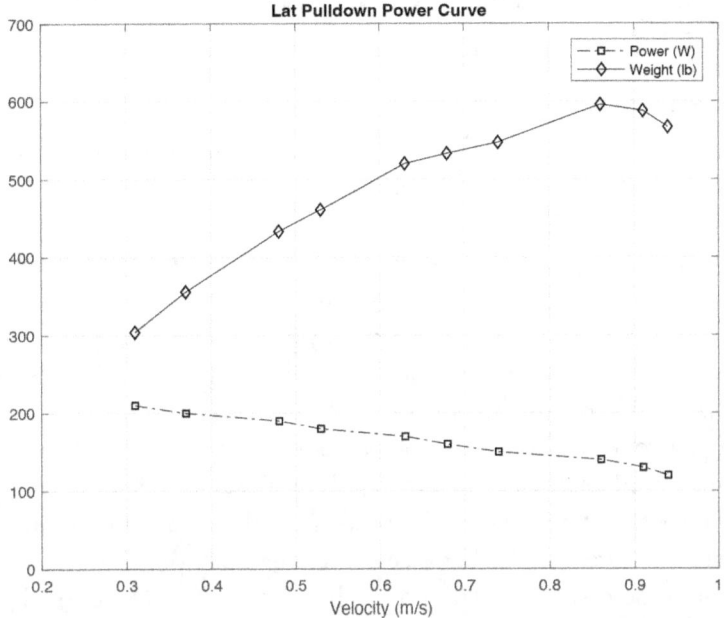

Figure 17.4: Lat pulldown power curve – power (Watts) and weight lifted (pounds) vs. average velocity (m/s).

A plot of average power for each set, i.e., at each weight, vs. average velocity, appears in Fig. 17.4. The abscissa (sometimes referred to colloquially as the "x-axis" or "horizontal axis") in this plot is the velocity in *m/s*. Consequently, my first sets, which resulted in the highest average velocity for the lowest weights, are on the right side of the plot. The slower average velocity data points on the left correspond to the later, higher-weight sets. I was working from right to left in the plot. The plot has a maximum at around 0.84 *m/s*, which I achieved using 140 *lb.* of weight plates. This was my average velocity for lat pulldowns that generates the maximum average power. From now on, to train for maximum power, I've targeted 0.84 *m/s* for this exercise. Based on the power curve I generated for cable rows, the target velocity is slightly higher for that exercise, 0.90 *m/s*.

My target velocity won't necessarily generate maximum power for anyone else. And it's not the velocity that generates maximum power for me on any other exercise, as noted above. By varying both velocity and weight I found my velocity sweet spot to maximize power for two exercises. And the literature suggests that while the weight I can move at

---

99 If your reps fall by more than 10% below your best rep's velocity during a set, you should stop the exercise and recover before continuing.

this velocity will increase over the coming months with training, the velocity at which I can generate maximum power should stay the same. That's the underlying benefit of VBT. You know what to target for each workout, and you know where you're at with each rep.

Note that my max power velocity of 0.84 *m/s* for lat pulldowns lies within the Strength-Speed range shown in Fig. 17.3. Lat pulldowns, overhead cable presses, inverted rows, and cable rows are functional, paddling-specific exercises. These define my circuit; you may find other exercises better suit your needs.

Caveats and Observations

Should you follow this path and try VBT? First off, you should consult your doctor before embarking on any exercise plan. Everybody runs a high risk of injury doing strength training, including VBT.

After completing compensation and base lifting phases during off-season training, I do three VBT sessions a week. I would not undertake VBT unless I had been doing strength training to some significant degree for at least a year or more. Why the note of caution? Because it's easy to injure shoulders and elbows, accelerating weight that fast. I'm very, very careful to pre-load each exercise. In other words, I engage the working muscle chain with the weight – or the grip for cable machines – before moving it. This prevents me from putting a jerk[100] load on my joints at the start of a rep. Jerking a load can be destructive to joints and connective tissue, both of which take a long time to heal. I'm careful to use excellent form. I strongly engage my core muscles before each set and keep the core engaged throughout the lift. And I terminate a set if any part of my body experiences even the slightest twinge.

Qualitatively, I've found that VBT has aided my starts and jumps. My peak speed during short interval workouts has increased. In other words, it has contributed to my ability to generate peak power.

Take-Aways

- Velocity Based Training (VBT) is a strength training methodology that prioritizes power development. VBT aligns nicely with paddling since the cruising speed of a hull is proportional to the paddling power you can supply.

- You now know how to measure your own power curves in the gym, and then design and implement a VBT program for your paddling season.

- Only undertake Velocity Based Training after establishing a solid base in conventional strength training. Otherwise you risk injury from the stresses placed on your muscles and connective tissues.

Addendum – Determining Resistance for Pushups and Inverted Rows

Using a velocity sensor requires that you enter a weight corresponding to each exercise to accurately compute power. The sensor does the velocity part. It may be tempting to use your body weight as input for pushups. But this is not correct. First, you have a single point of

---

100 Jerk is the time rate of acceleration. When jerk is greater than zero your acceleration is accelerating.

contact (your feet) between you and the ground when you're standing. That's why a weight scale works: there is no other path for weight to "pass through" to the ground. But when you do a pushup, you contact the ground through your feet and your hands. There is a force transmitted through all points of contact. Since the total applied force must equal your body weight, the force you move in executing a pushup must be less than your total weight. This is because your feet serve as pivot points that bear weight but don't translate up or down.

How do you determine the weight that moves during a pushup? An engineer might construct a free body diagram and impose a static equilibrium requirement. The easier way is to take out a weight scale, get on the floor like you're going to do a pushup, and put both of your hands atop the scale. Take the reading: that's the weight you move with each pushup. You also have two force transmission points for inverted rows: through your heels to the ground and through your hands to a bar above you. Since the inverted row is essentially an upside-down pushup, the weight you move during each repetition is the same as you move to do a pushup. You can prove it yourself by drawing a free body diagram.

Further Reading

Bryan Mann, *Developing Explosive Athletes: Use of Velocity Based Training in Training Athletes*, 3rd Edition, Ultimate Athlete Concepts, Michigan USA (2016).

Tudor Bompa, *Periodization*, and *Periodization Training for Sport*, Human Kinetics, Champaign, IL (1999).

CHAPTER 18

The Deflection Point

Introduction: Pain with No Gain

I'm sure some of you have experienced this at a race: You go out hard at the start, trying to hang with the leaders. Then, maybe thirty to forty-five minutes into the race, your body says, "No more!" Crash and burn. Going in you felt well trained, rested, adequately nourished, hydrated, calm, and focused coming into the event. But as you slowly cruise into the finish, you wonder what the heck just happened.

While there are any number of reasons for this kind of performance – and I've experienced them all – you may have pushed too hard at an anaerobic pace beyond your current fitness level. Your body couldn't keep up with the production of blood lactate, hydrogen ions, and other muscle metabolism by-products. Remember what we covered in Chapter 16. Your body does all that stuff. Not only did you crash and burn, you experienced science!

As we saw in Chapter 16, the various energy systems – aerobic, anaerobic, and phosphate – are called upon in proportion to exercise duration and intensity. The body knows perfectly well when it is performing aerobically or anaerobically. How do *we* know whether our workouts are aerobic, anaerobic, or somewhere in between? And can we use this to determine how hard to paddle during a race?

This chapter will present a test method that lets you gauge whether your body is performing aerobically or anaerobically. We'll show you how to use a paddle ergometer[101]

---

101 And possibly, on the water, with various qualifiers and course requirements.

and a heart rate monitor to determine experimentally a transition point called the *lactate threshold*, above which your body cannot keep up with the production of lactate. Think of this threshold as the red line on your car's tachometer. Once you surpass and linger above the red line, the clock starts ticking until you flame out.

We'll use heart rate to quantify exercise intensity – the harder you work, the faster your heart beats. For running, the relationship between heart rate and exercise intensity is linear up to the lactate threshold. This means that as speed increases, your heart rate rises in linear proportion to speed.

But as we learned in Chapter 1, the drag force on a hull is proportional to the square of speed. Drag force is not a linear effect. Consequently, the work you expend paddling a hull increases nonlinearly with speed, too. Exercise intensity not only grows, but it grows exponentially. This nonlinearity complicates how we determine lactate threshold since most methods assume a linear relationship between heart rate and exercise intensity. This is where the physics of paddling comes to the rescue.

Since many readers may be unfamiliar with the minutia of heart rate data and blood lactate concentration, we'll review these topics in the following two sections. Then, we'll use these tools to contextualize heart rate data obtained in a paddling "step test." You'll get to be a paddling scientist in the comfort of your gym.

Heart Rate

Our circulatory system transports oxygen and glucose to muscles during exercise and removes exercise's byproducts, including heat, carbon dioxide, water, and hydrogen ions. During exercise, the capillaries expand, enabling the heart to deliver more oxygenated blood to the muscles.[102] At rest, the heart pumps about 5 liters of blood per minute. As you begin to exercise, the autonomic nervous system tells the heart to beat faster. At maximum output, the heart of a highly trained athlete can pump as much as 40 liters of blood per minute.

As noted above, there is a relationship between heart rate and exercise. Consequently, heart rate is commonly used to measure exercise intensity, and for conducting a targeted training program. Our maximum and resting heart rates are of particular interest as these prescribe training zones, track the progress of your exercise program, and help us monitor recovery from workouts.

Everyone has a *maximum heart rate* (HRmax) that does not change appreciably with endurance training. The maximum heart rate does, however, decrease with age. It is challenging to determine HRmax without directly measuring it. However, we can approximate HRmax via

$$\text{HRmax} \sim 205 - (\text{age} \div 2) \text{ for men,}$$
$$\text{HRmax} \sim 210 - (\text{age} \div 2) \text{ for women.}$$

According to this formula, a 44-year-old woman has a maximum heart rate of 210 − (44÷2),

---

102 At the same time, the brainstem is sending signals that constrict the flow of blood to the internal organs, enabling more blood to flow to the muscles.

or 188 beats per minute ('*bpm*'). Note that an athlete's actual maximum heart rate can differ significantly from what is predicted by this simple formula. For accurate heart rate-based training, you can measure HRmax in a threshold test or race.

A chest strap-based heart rate monitor is a convenient tool for measuring heart rate.[103] These devices consist of a sensor, a transmitter, a receiver, a microcontroller, and a display. Sensing electrodes are embedded in the chest strap. These electrodes measure electrical activity when nerves fire to make the heart beat. A transmitter in the chest strap sends this data to the receiver using a low-power radio signal. The receiver is built into a watch housing and worn on the wrist, mounted on a thwart, bulkhead, or the like. The wrist unit displays heart rate. More advanced models also measure speed and distance using other sensors such as GPS receivers or accelerometers.

You can accurately measure HRmax in the field using a heart rate monitor. Here are two methods:[104]

- THRESHOLD TEST: First, you must recover fully from previous anaerobic workouts or races and be well-rested. The test begins with a warmup of 15 to 20 minutes, with gradually increasing intensity. This is followed by an all-out effort of 20 minutes. The level of effort over the final 4 to 5 minutes should be maximal, with the final 30-60 seconds an all-out sprint. You should feel that there is nothing left in the tank at the end. Over this final 30 seconds the heart rate will be very close to HRmax, usually within two beats per minute or less.

- RACE TEST: Finish a 5- to 10-kilometer race with a kick over the final 4 to 5 minutes. The last 30-60 seconds should be an all-out sprint, and you should feel that there is nothing left in the tank at the finish line. The heart rate over this final 30 seconds will be very close to HRmax.

An important caveat about HRmax is that it is *sport specific*. HRmax for running will be higher than HRmax for paddling, cycling, or swimming; differences of 10 *bpm* to 15 *bpm* are common. This variation is because of the differing mix of muscles engaged during running, paddling, swimming, cycling, etc. For example, one model suggests a paddler uses 85% of the aggregate muscle engagement of a runner. This is because the hull supports a paddler's body against the pull of gravity; paddling is a "weight supported" sport. The paddler's legs only perform isometric contractions during the leg drive. Since HRmax is used to plan training, if you are a multi-sport athlete, it can be measured for each sport using the methods outlined above.

So why do we care about HRmax? Because every paddler's HRmax will be different, for starters. I'm always amused at races when I hear someone say, "My heart rate got up to 175!" Well, OK, is that good or bad? Does it mean they were working hard or not? Unless

---

103 For those of you who rely on wrist-mounted optical / IR heart rate sensors, note that these can produce inconsistent data depending on how they are worn. Your experience may differ but consider using a chest strap if you can, or an IR sensor worn on the forearm.

104 Be sure to consult with your physician before undertaking any exercise program and before performing any of the performance tests outlined in this book!

we know that person's maximum heart rate, the number is mostly meaningless. Further, since most heart rate-based training prescriptions are expressed as a percentage of HRmax, it is an important reference.

Figs. 18.1 and 18.2 are sample heart rate records plotted over time. The first shows heart rate as a percentage of HRmax over a 55-minute race. Nearly steady-state heart rate is reached in about 3 minutes, the time for the aerobic system to engage fully. There was a buoy turn at the 19-minute mark – note the increase in heart rate as the paddler makes the turn and briefly accelerates. Otherwise, the heart rate is relatively steady, reflecting a constant effort throughout the event. Heart rate rises during the finishing kick, reaching 98% HRmax at the end. The steady-state heart rate – 89% HRmax – was likely close to this paddler's lactate threshold heart rate. And as we'll learn below, that's a very good thing for a race of this length.

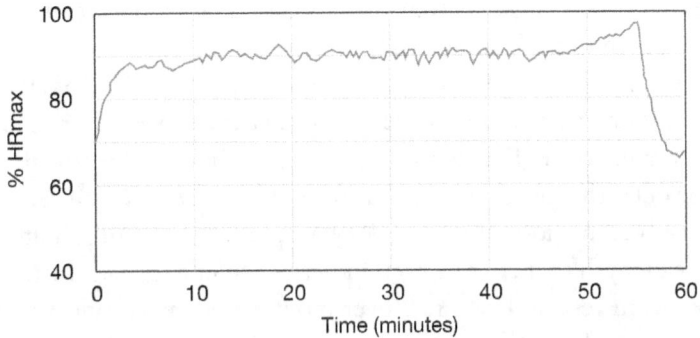

Figure 18.1: Heart rate plot for a 55-minute race.

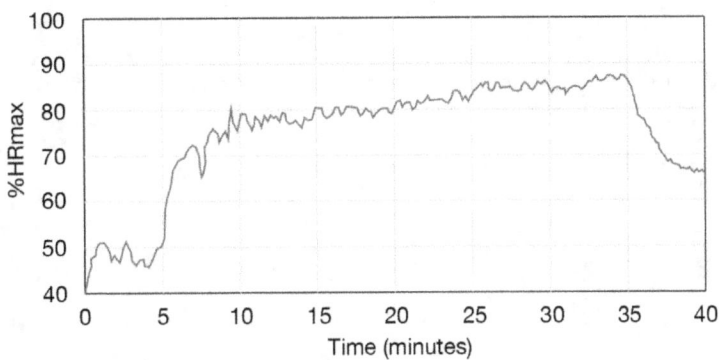

Figure 18.2: Example of heart rate drift.

Fig. 18.2 plots heart rate as a percentage of HRmax for a canoeist paddling at a constant pace. Notice how heart rate increases over time during the work interval. This phenomenon is called *heart rate drift*. In fit athletes, heart rate drift is attributable to core temperature increase. Or, in long races, it can be attributed to dehydration. The heart must pump harder with less blood volume to supply the working muscles with oxygen and remove waste products. In long races, heart rate can even drift downwards when paddling at a constant level of effort.

Trained endurance athletes tend to have a lower *resting heart rate* (HRrest) than untrained individuals. The average person's resting heart rate is between 70 *bpm* and 80 *bpm*. As fitness improves, HRrest gradually decreases. A well-trained athlete can have a resting heart rate as low as 40 *bpm* to 50 *bpm*. This is due in part to conditioning-based changes in the heart. Endurance training increases the heart's stroke volume; it can pump more blood per beat. Also, the working muscles' metabolic systems become more efficient with training, requiring less oxygen at rest.

You can measure resting heart rate in the morning, right after you wake up. Studies have shown that HRrest is generally lower in the morning than in the evening.[105] You can use your heart rate monitor to measure HRrest or an inexpensive pulse oximeter that also displays your pulse. However, manually measuring the radial pulse will suffice.[106]

Some people count their pulse over ten seconds and multiply the result by six to get their heart rate in beats per minute. Taking your pulse over a full minute will increase accuracy; if you miscount your pulse over 10 seconds by one beat, your estimate of HRrest will be off by 6 *bpm*. Are you really in such a hurry that you need the extra 50 seconds in your day? As you measure heart rate, notice that your pulse increases when you breathe in then decreases when you breathe out. This phenomenon, called *heart rate variability*, (generally) reflects a healthy heart. Just breathe normally and take your pulse over a full minute.

It is a widespread myth that HRrest itself indicates athletic fitness. Instead, gradual and systematic decreases in resting heart rate over months and years reflect improved conditioning. More importantly, HRrest is an important indicator of recovery from the previous day's workout or race. If the resting heart rate is elevated, for example, by 5 *bpm* compared to your usual baseline, then recovery may be incomplete. You might consider substituting a recovery workout for that day's training. The morning pulse should recover to baseline by

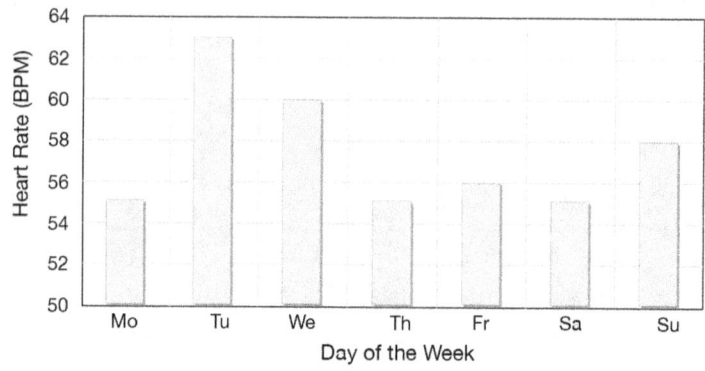

Figure 18.3: HRrest over time.

---

105 If your resting heart rate is consistently lower in the evening, however, measure it then.

106 Your heart rate will elevate by around 2-3 *bpm* if you need to go to the bathroom. So, if you need to pee when you wake up, take care of business, then climb back into bed. Your heart rate will settle back down in a few minutes, and you can get a more accurate HRrest reading.

the next day. Increases in HRrest that do not subside can be an early indication of over-training, stress, or looming viral infections. In that case, it may be relevant also to measure changes in blood oxygenation or see your doctor.

Fig. 18.3 shows an example of how the resting heart rate can change over time. This is a plot of my morning rest heart rate over one week. When I took this data, my normal resting heart rate was 54 *bpm*. Note how HRrest rose by eight bpm when I fought a minor cold on Tuesday and Wednesday, then returned to normal levels after it had passed. Sunday's elevated resting heart rate reflects Saturday's hard 2-hour workout.

Blood Lactate

As you exercise, your glycolytic system produces lactic acid – see Chapter 16 for details. Before a training session, most paddlers have a blood lactate concentration of approximately 1.5 millimoles per liter (*mmol/l*). Once you start paddling aerobically, the concentration drops below 1.0 *mmol/l*. As you continue paddling, it will rise depending on your level of effort.

The relationship between blood lactate production and exercise intensity has linear and nonlinear components, as shown conceptually in Fig. 18.4. For running, this dependence is fairly linear below a lactate concentration of 1.5 *mmol/l*, then starts to bend sharply upward as the concentration approaches 4.0 *mmol/l*. The 4.0 *mmol/l* blood lactate concentration point occurs at the transition between aerobic and anaerobic metabolism. You can train or race continuously at the lactate threshold for about an hour before flaming out; less if you race harder; longer if you periodically drop down below the lactate threshold to recover – like dropping in to ride a boat's stern wake.

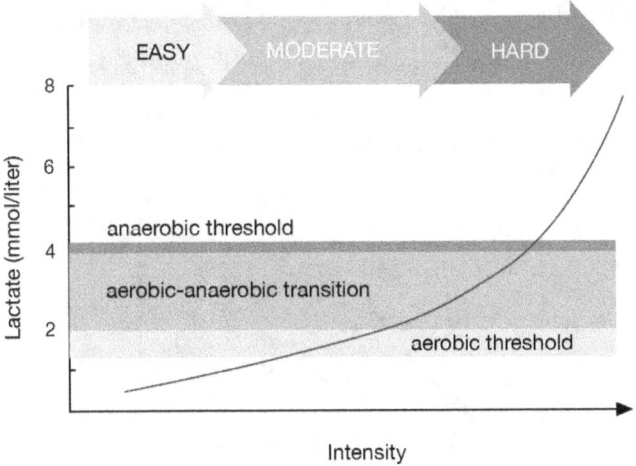

Figure 18.4: Blood lactate concentration as a function of exercise intensity.

The blood lactate vs. exercise intensity curve will shift to the right as you become more fit. With increasing fitness, your aerobic system will clear the metabolic byproducts of exercise more efficiently. The glycolytic system and the Cori cycle will become more efficient, especially after anaerobic workouts such as tempo paddles and racing. With this kind of training, you'll be able to paddle harder for more extended periods because your lactate threshold shifts to increasingly higher exercise intensities.

This *training effect* makes determining metabolic training zones problematic since their boundaries shift with your changing fitness level. The precise solution is to periodically measure lactate concentration from blood samples taken at various workout intensities. We can measure blood lactate using a drop of blood from a fingertip or ear lobe much the same way diabetics monitor the level of blood sugar. Portable blood lactate analyzers enable paddlers and coaches to accurately determine blood lactate levels in the field to precisely determine individualized metabolic training zones. They are the gold standard. While not practical for most of us, they may work for a paddling club or team.

The Deflection Point and Lactate Threshold

But what about the rest of us? Can we leverage biology and mathematics to determine the transition from aerobic to anaerobic paddling? Recall that the lactate concentration vs. exercise intensity curve, Fig. 18.4, has a "knee." And we also know that paddling a nonlinear relationship between heart rate and exercise intensity owing to drag's nonlinear dependence on speed. With a bit of tinkering, we can take advantage of these facts to determine non-invasively the 4 *mmol/l* blood lactate concentration breakpoint and its associated heart rate.

The *heart rate deflection point* (HRdefl) is the heart rate above which the accumulation of lactate exceeds the liver's ability to recycle it. This corresponds to the lactate threshold of 4.0 *mmol/l*. A fit paddler can perform at the intensity level corresponding to HRdefl for about an hour since there is equilibrium between lactate production and removal. Once you know your lactate threshold heart rate, you can determine your lactate threshold pace by paddling at HRdefl and noting your pace using a GPS sports watch. Most importantly, we can increase the pace corresponding to the deflection point through training.

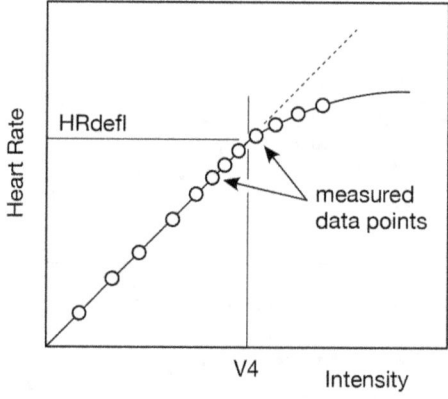

Figure 18.5: Idealized Conconi test plot for running.

Italian physiology professor Francesco Conconi developed a means to measure the lactate deflection point non-invasively. Otherwise, we must take blood samples to determine the heart rate and exercise intensity at which blood lactate concentration reaches the lactate threshold. A so-called *Conconi Test* maps heart rate vs. exercise intensity at several intensity levels, i.e., at several pace/speed set points. The heart rate vs. pace plot will initially have a generally linear trend for runners, as shown in Fig. 18.5. The heart rate versus intensity plot curves downward at the deflection point. The body's energy systems, which deliver oxygen to and remove lactate from the working muscles, can no longer "keep up" at higher intensities. The dominant energy mechanism transitions from aerobic to anaerobic. This knee in the curve corresponds to the 4.0 *mmol/l* blood lactate threshold.

Figure 18.6: Paddle ergometer with a performance monitor.

You can conduct a Conconi test on a paddling ergometer equipped with a performance meter like that shown in Fig. 18.6. You'll need a heart rate monitor, a sports computer that you can program to direct custom workouts, software to average heart rate data, and software like Microsoft Excel for plotting and basic mathematical computations. It would be best if you also were fully recovered from any race, interval workout, long paddling session, etc., before you start. Track your resting heart rate or heart rate variability to be sure.

The process is straightforward. That said, your first test will be a learning experience as you experiment with the ergometer's damper setting, starting pace for the first testing step, and the number of steps. Once you have all these parameters sorted, you can repeat the test every 4-6 weeks to assess improvements in the anaerobic threshold from your training. Note that the test must be conducted under identical conditions to be a valid indicator of training progress.

If you are using a Concept2 rower with a paddling adapter, set the damper to 2 or 3; the precise setting will take a bit of experimentation. An extensive warm-up is followed by a "step test." During this step test, you decrease the pace[107] at regular intervals as defined in Fig. 18.7. Program your sports watch to do an interval workout[108] comprising:

1. A 20-minute warmup where you gradually increase the intensity, followed by
2. Ten to twelve two-minute segments at a target pace (and no recovery phase), then
3. A five-minute cool-down.

The warmup step is performed at a pace corresponding to 55-60% HRmax; for me, this was 3:00/500m. Your pace may be different depending on your physiology, fitness, and damper setting. Then, after reaching each step's target pace, hold that pace constant for two minutes. This requires a fair bit of focus and concentration, especially over the last few steps. Record your heart rate over the entire test with a heart rate monitor.

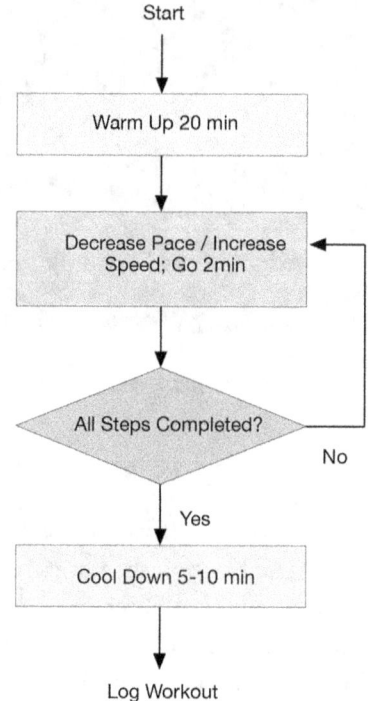

Figure 18.7: Conconi Test logic.

The pulse rate over the last minute of the step is the steady-state heart rate for that pace. The heart rate needs time to stabilize at the level of effort, as seen in Fig. 18.8. Note how the heart rate rises over the first minute, then is relatively constant through the balance of the two-minute segment. You can either average the data over the last minute or the entire

---

107 Which means the speed increases; pace is speed's inverse.

108 Sports watches that have an "auto lap" feature based on time are ideal since they will break each step into a separate lap, and you can extract and average the heart rate for each of these segments separately.

segment; just be consistent. Your estimated lactate threshold heart rate may be a couple of beats per minute low if you choose the latter.

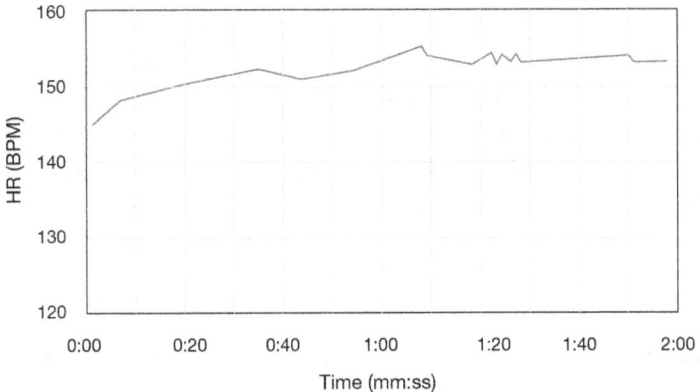

Figure 18.8: Heart rate drift over a 2-minute segment.

Incrementally decrease pace by 5*sec*/500*m* for each segment. Record data points at and above the anaerobic threshold, which typically occurs around 85-95% HRmax.

I noted something odd when I plotted average heart rate data vs. step from my testing. The data is plotted in Fig. 18.9. The heart rate values don't fall on a straight line before bending to the right. Instead, they exhibit a bit of a curve there. For comparison, I plotted one dashed line where I fit a linear trend line to some of the upper points and another one where I searched for a linear dependence among the lower points. While you might argue that the curve "deflects" at the 8th segment, you'd be hard-pressed to justify that choice. Why not the 7th? Or the 9th? Also, why does the plot bend over the lower eight segments?

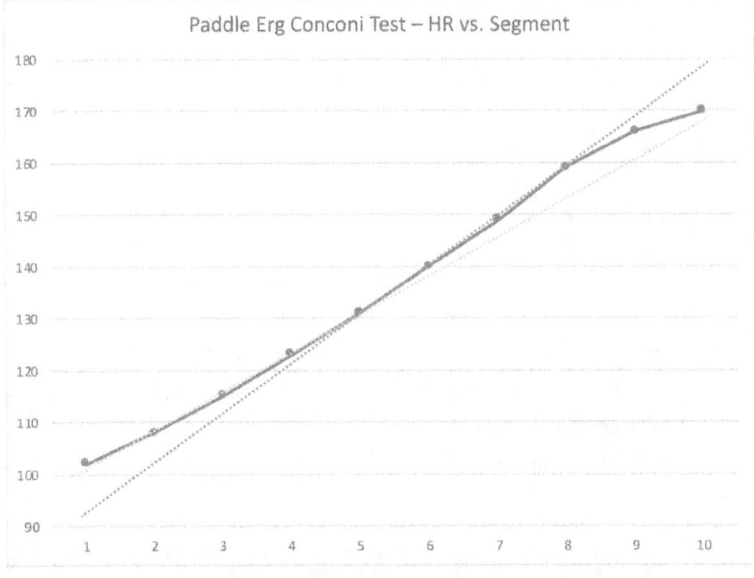

Figure 18.9: Heart Rate vs. Segment (solid trace).

From Chapter 1, we know the drag force varies as the square of the velocity at all but the lowest speeds. The paddle ergometer mimics this behavior. Consequently, the heart rate versus intensity plot should also display that kind of nonlinear behavior, even below the deflection point. We just need to unravel the quadratic dependence.

This is where high school algebra helps us. As you may recall, there is a transcendental function called the logarithm. Napier invented logarithms back in the 1800s to help compute the products of large numbers, among other things. Since then, other interesting properties of logarithms have been deduced. Taking the logarithm of any function expressed as the power of a variable reduces to a simple linear multiplicative form:

$$\log\left(x^n\right) = n\log\left(x\right). \tag{18.1}$$

That is, something that embodies a quadratic dependence on a variable now takes on a linear form if we plot its logarithm.[109]

Armed with this insight, I computed the natural logarithm ("ln" or "$\log_e$") of my heart rate data and replotted it. The result is shown in Fig. 18.10.

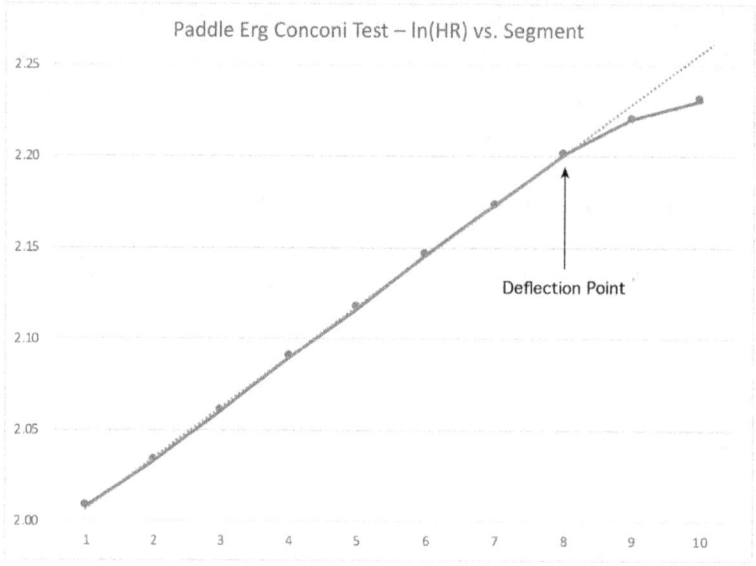

Figure 18.10: Loge(Heart Rate) vs. Segment.

I added a straight-line fit to data points one through eight; the fit is surprisingly good. The deflection point is clear, occurring at or just above segment 8's heart rate. Segment 8 corresponded to an average heart rate of 159 *bpm*, or 88% HRmax for me. As you can see, physics made the heart rate data analysis much more manageable.[110]

---

109 This is also true for cubic, quartic, or any other power for that matter, including fractional powers. For those of you keeping score at home, this power-law-to-linear transformation is independent of the logarithm's base. For the analysis here, we'll use the natural logarithms, which are base *e*.

110 I was stunned at how good the data came out. So, I repeated the test a week later, and it was just as good.

These test results don't map to pace-based performance on the water since they are done on an ergometer using heart rate. But they do map well based on heart rate since the ergometer does a good job mimicking a hull's resistance and paddling motion. Once you know your in-boat pace at HRdefl, you can bound race performance as well. You'll know where the "red line" is on your internal tachometer and can adjust your level of effort accordingly.

You can conduct a Conconi test on the water if you carefully control environmental factors. You'll need a data recording GPS/heart rate monitor that lets you program the sequence of work segments defined in Fig. 18.7. As you paddle each step, the sports computer records distance traveled, pace, and heart rate. Use averaged data and perform the analysis as outlined above. You should paddle each step on a long deep-water course without wind or current. Otherwise, shallow water effects will corrupt your data; wind or current will bias pace data.

Regrettably, the Conconi test doesn't work for every paddler. Some paddlers perform a step test, and the resulting data doesn't reveal an inflection point. While this may reflect an insufficient warm-up, insufficient pre-test recovery, or step lengths that are too short, some paddlers have an indistinct deflection point. If this happens to you, try the test again in a couple of weeks, paying careful attention to pre-test recovery, proper warm-up, and step length. If the resulting test data is still ambiguous, consider using a blood lactate test to determine your anaerobic threshold. Or use time trial results to estimate HRdefl. Recall that you can paddle at your anaerobic threshold heart rate for about an hour, so find a suitable race or course offering that duration, paddle while holding a consistent heart rate and level of effort and listen to your body – and write down that heart rate!

Take-Aways

- Lactate threshold plays a significant role in determining your achievable race pace.
- For running, the relationship between heart rate and exercise intensity is linear up to the lactate threshold. This relationship has a breakpoint at the threshold. You can exploit this feature to determine race pace by plotting heart race versus speed data.
- When paddling, exercise intensity and speed are not linearly related owing to drag's quadratic dependence on speed. It can be challenging to pinpoint the breakpoint in a plot of heart rate versus paddling speed.
- Plotting the logarithm of heart race versus speed untangles the nonlinear effect of hull drag, so you can experimentally determine your paddling-specific lactate threshold.
- Some paddlers' heart rate versus speed curve doesn't have a clearly demarked breakpoint. For them a direct measurement of blood lactate may be needed.

Your mileage may vary.

Extra Credit: A Note About Oxygen

Some of you may think that lactate threshold sounds interesting, but isn't VO2max more important? Important, yes. But *more* important?

Aerobic metabolism uses oxygen to fuel the electron transport system. Oxygen binds to an iron-rich protein in the blood called hemoglobin and is transported from the lungs to the muscles. Oxygen unbinds from the hemoglobin to feed the various muscle oxidative sites. Then the hemoglobin binds to carbon dioxide produced by muscle metabolism. This carbon dioxide is carried to the lungs and expelled when you exhale. The volume rate of oxygen you can process metabolically is often written as VO2 ("Volume of O2"), and the maximum volume rate of oxygen you can process is called VO2max.

VO2max is the most common measure of maximal aerobic capacity. It depends on oxygen diffusion in the lungs, hemoglobin binding efficiency, blood transport, and mitochondrial density and efficiency. It also depends on genetics. Athletes reach VO2max when their oxygen consumption stops increasing while physical exertion continues to increase. The body needs to produce approximately ten *mmol/l* of blood lactate to reach VO2max, and a respiratory quotient – the ratio of carbon dioxide output to oxygen consumed, denoted as 'RQ' – above 1.1.

VO2max can increase with training by up to 20 percent. After that, for many athletes VO2max only increases slightly with additional aerobic training. Further, aerobically fit athletes show minimal fluctuation in VO2max over a season.

You will see the most significant improvements in maximum oxygen uptake by exercising at intensities between 90 and 100 percent VO2max. A 90-100% VO2max effort roughly corresponds to exercise intensities of around 90-100% HRmax. This range corresponds to sprint interval and high-intensity continuous training, an intensity seen in dragon boat paddlers and sprint canoeists and kayakers.

Since VO2max occurs at approximately ten *mmol/l* blood lactate concentration, you will only be able to perform at this intensity for a few minutes. For longer races, your performance potential depends far more on the submaximal percentage of VO2max that you can sustain. This is where the anaerobic threshold factors in.

Think of VO2max as an indicator of an athlete's aerobic potential; the anaerobic threshold indicates an athlete's utilizable aerobic capacity. For example, a well-trained athlete's oxygen consumption at the anaerobic threshold is almost 90% of VO2max; in untrained individuals, the anaerobic threshold may occur closer to 60% of VO2max. Race times correlate very well with blood lactate production. This highlights the importance of aerobic utilization and the lactate threshold.

Further Reading

Burke, Edmund, *Precision Heart Rate Training*, Human Kinetics, Champaign, IL (1998).

Janssen, Peter, *Lactate Threshold Training*, Human Kinetics, Champaign, IL (2001).

Nolte, Volker, *Rowing Faster*, Human Kinetics, Champaign, IL (2005).

Sleamaker, Rob, and Browning, Ray, *SERIOUS Training for Endurance Athletes, 2nd Edition*, Human Kinetics, Champaign, IL (1996).

P. Droghetti, C. Borsetto, I. Casoni, M. Cellini, M. Ferrari, A.R. Paolini, P.G. Ziglio, and F. Conconi, "Noninvasive determination of the anaerobic threshold in canoeing, cross-country skiing, cycling, roller and ice skating, rowing, and walking," *European Journal of Applied Physiology* **53**:299-303 (1985).

CHAPTER 19

Paddling 30-30 Intervals

Introduction

Elite paddlers have excellent biomechanics, a high lactate threshold and lactate threshold pace, the ability to get power in the water, and a high physiological capacity to process oxygen during races. We characterize oxygen uptake by VO2max, the maximum oxygen volume you can process during aerobic activity. All of these factors are trainable. While it's tempting to consider each in isolation, these factors are interrelated. For example, improving biomechanical efficiency leads to improved VO2max without contributing to the number and density of mitochondria in the working muscles.

Paddling VO2max is tied to speed since the maximal oxygen consumption threshold occurs at a certain speed.[111] Consequently, one predictor of race performance is the velocity at VO2max, commonly written as "vVO2max."[112] This is the lowest sustainable speed where you perform at the maximum oxygen consumption level of effort. For example, if

---

[111] We consume oxygen as part of our metabolism. While we process oxygen when we're at rest, we process more oxygen when active. Oxygen uptake depends upon the muscles that are engaged in an activity and the degree to which they are engaged. Paddling utilizes different muscles than running. As you paddle faster, you consume a greater volume of oxygen until you reach a level of effort and speed where your body cannot process more. This is your paddling VO2max. Hence VO2max depends on speed as well as the activity (paddling vs. running, etc.).

[112] Recall that velocity is a vector; what we mean here is speed, which is a scalar. But vVO2max is a term of art, so we'll use it.

you reach your maximum oxygen consumption rate at 7 *mph*, and oxygen uptake does not increase as you go faster, then your vVO2max is 7 *mph*.

Improving VO2max increases your vVO2 max, as well as the time you can sustain this pace. Since VO2max is trainable, what is the best way to increase it? VO2max depends upon improvements in biomechanics and aerobic capacity, so one approach is to paddle more using excellent technique. Another way is to train at vVO2max. But paddling at vVO2max is only sustainable for 6 to 11 minutes – this type of workout is taxing. Is there a way to bank more time at VO2max without undue recovery time or possibly overtraining?

At this point, many readers will say, "Do intervals!" And they're right. But is there an optimal interval workout that maximizes time at VO2max? And can it be adapted to paddlers?

In this chapter, we'll consider "Billat's 30-30" workout to achieve this. this workout was initially developed for runners. We'll use physics to adapt her protocol based on percentages of vVO2max. Along the way, we'll see why paddlers should once again account for the differing relationships between power and speed for running vs. paddling.

The 30-30 Workout

The Billat 30-30 is named after its creator, Veronique Billat, an exercise physiologist at the University of Ille. Billat studied workouts where runners spend the largest total amount of time at VO2max, presumably leading to the greatest increase in VO2max. She and her colleagues found that runners seeking to improve maximum oxygen uptake should run no faster than vVO2max since they fatigued quickly at faster speeds. They also found that a runner's oxygen consumption remains at or near VO2max for up to 15-20 seconds *after* they stop running at vVO2 max pace. A workout designed to exploit this lag would allow a runner's cardiovascular system to spend more time at VO2max while spending less time running there. Maximum oxygen uptake improves with less physical wear and tear.

This prescription naturally lends itself to intervals performed at vVO2 max with short recoveries at a slower speed. Billat and her colleagues set this recovery speed at 50% of vVO2max. Their workout consists of 30-second work phases at vVO2max separated by 30-second recoveries at half the work phase speed. This sequence is repeated to failure, e.g., until vVO2max can no longer be sustained for 30 seconds. Keeping the interval work phases short delays muscle fatigue. Short recoveries keep the body effectively working at or near VO2max while operating at a slower and more muscle-friendly pace.

Some runners were able to spend more than 18 total minutes at VO2max in a single workout using this format, with much of that time accrued during the 50% vVO2max recoveries. A group of moderately fit runners increased their VO2max by 10 percent over 8-10 weeks while undertaking twice weekly 30-30 sessions. For competitive athletes, a 10% increase in VO2max is impactful.

Time for Some Science

Running and paddling differ not only in muscle engagement but in the relationship between speed and applied power. This might lead you to ask, "Is half of vVO2max the same for runners and paddlers?" In other words, if I perform a paddling interval workout with work phases done at vVO2max,[113] then halve my paddling speed during the recovery phase, will I get the same benefit as a runner who halves their running speed during recovery? In other words, will I match Billat's 30-30 protocol *physiologically* if I halve my speed on the paddle erg?[114]

What follows is a modest proposal, based on a syllogism and a little bit of physics. To motivate the syllogism, we'll consider the relationship between walking / running speed and oxygen consumption. This is depicted in Fig. 19.1 for a *70kg* athlete based on an American College of Sports Medicine (ACSM) model:

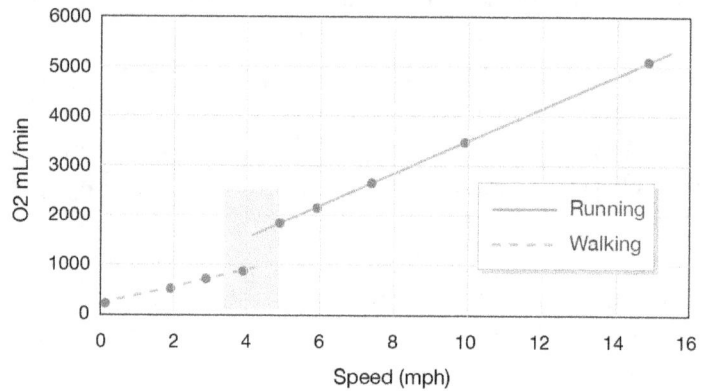

Figure 19.1: Model of running O2 consumption rate vs. speed.

Note that there is a breakpoint in oxygen (O2) consumption between walking and running. The biomechanics and muscle engagement differ between these two modes of movement.[115] Also, this model doesn't incorporate the plateau in O2 consumption that occurs at vVO2max. That said, the relevant takeaway here is that the relationship between O2 consumption and speed for running over a range of running speeds is *linear*.

Next, consider the relationship between walking / running speed and applied power, depicted in Fig. 19.2 for a *70kg* athlete using the same ACSM model. Note once again the breakpoint in this power plot between walking and running. Like Fig. 19.1's plot of running O2 consumption vs. speed, the relationship between applied power and speed for running over a range of running speeds is linear. One can infer that there is a correlation between O2 consumption and power expenditure for running and that both *linearly* depend on running speed.

---

113 We'll suggest ways to approximate vVO2max based on a simple time trial. Or you can go to a sports testing lab.

114 Note that this question is different than, "Will I get the same improvement in VO2max if I just halve my speed?" To this, I have no answer; there's not enough data. In this chapter, we'll limit ourselves to looking for ways to match Billat's protocol. This is an important distinction.

115 There is a similar breakpoint in the relationship of foot contact time and pace between walking and running, as we saw at the beginning of Chapter 7. The relationship describes a straight line for each way of moving, but their respective slopes are different.

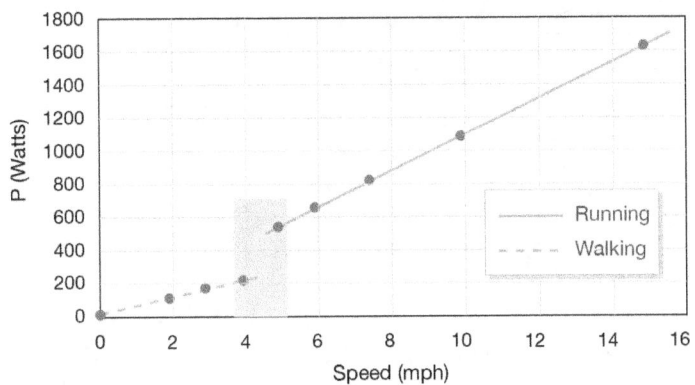

Figure 19.2: Model of running Power vs. speed.

From Chapter 1, we know that for paddling the relationship between paddling power P and hull speed v_{hull} is *cubic* rather than linear:

$$v_{hull} \propto \sqrt[3]{P} \;\rightarrow\; P \propto v_{hull}^3 , \tag{19.1}$$

where the truncated infinity symbol means "proportional to." Since we produce paddling power metabolically, we infer that paddling O2 consumption will have this cubic relationship to paddling speed. We need to account for the cubic relationship between power and speed to determine the paddling speed that halves applied paddling power. This means for half the paddling power P,

$$\sqrt[3]{\frac{P}{2}} \propto 0.794 \cdot v_{hull} , \tag{19.2}$$

where 0.794 is the cube root of one half, rounded to three digits. Considering the relevant physics, Billat's "50% vVO2max" *for paddling* is instead 79.4% vVO2max. To perform a paddling version of Billat 30-30 interval workout, we alternate between 30-second work phases at vVO2max and 30-second recovery phases at 79.4% vVO2max. Or if you like round numbers, 80% vVO2max; it's close enough.

For those of you who have done paddling intervals, this makes sense. Going from a work phase at 7 *mph* to a recovery phase at 3.5 *mph* – half of 7 *mph* – feels like you're hardly working at all. The goal of Billat's 30-30 is to allow the muscles to recover enough while keeping the body working metabolically at or near its maximum oxygen uptake. By contrast, going from 7 *mph* to 5.6 *mph* – 80% of 7 *mph* – feels qualitatively about half as hard.

Table 1 presents a list of possible vVO2max speeds/paces and their corresponding 80% speeds and paces. Paces for those who train on Concept2 ergometers – or who prefer MKS units – are included since these commonly display pace in units of *min/500m*.

TABLE 19.1: SPEED, PACE, AND 80%.

speed (mph)	pace (min/mi)	pace (min/500m)	80% speed (mph)	80% pace (min/mi)	80% pace (min/500m)
6.0	10:00	3:06	4.76	12:38	3:55
6.5	9:13	2:51	5.16	11:37	3:36
7.0	8:34	2:39	5.56	10:48	3:21
7.5	8:00	2:29	5.95	10:05	3:07
8.0	7:30	2:19	6.35	9:26	2:55

A 30-30 Interval Protocol for Paddlers

What then is the 30-30 workout protocol, and how do you determine your vVO2max? First, the only sure way to determine VO2max and vVO2max is in a properly outfitted lab using a gas analyzer. Lacking that, you can get a good approximation by paddling a maximum effort time trial lasting between 6 and 8 minutes – after a thorough warmup, of course.[116] The speed you can sustain over that time trial will be reasonably close to vVO2max. You'll get feedback from the 30-30 interval workout itself that will help tune your vVO2max speed in the future.

The 30-30 workout for paddlers is then:

1. Perform a thorough warmup with 15-20 minutes of easy paddling
2. Paddle at vVO2max for 30 seconds (work phase)
3. Paddle at 80% of vVO2max for 30 seconds (recovery phase)
4. Repeat steps 2 and 3 until you can no longer maintain vVO2max during the work phase.
5. Cool down with easy paddling for at least 10 minutes

Modern sports watches can be programmed to provide timed alerts at 30-second intervals. A GPS sports watch will help you set the pace if you're doing the workout on the water; a performance monitor will help if you're using a paddling ergometer.

How many intervals should you be able to perform? Somewhere between twelve and twenty. If you can do more than this, your estimated vVO2max is too low. It would be best to increase speed during the work phases until you can do around twelve to sixteen intervals and no more. And if you can't do twelve, no problem. Just dial back vVO2max a bit for the next 30-30 workout. Over time the number of intervals you can do will increase, as will your vVO2max. You can adjust the protocol accordingly. Since this is a High-Intensity Interval Training (HIIT) workout, most of us will do it no more than once per week, typically during a training season's "building" phase. Be sure to monitor your workout recovery to prevent overtraining. Continue doing the paddling Billat 30-30 workout once a week for, say, three months or until you hit your transition phase and race season.

---

116 Be sure to consult with your physician before undertaking any exercise program and before performing any of the performance tests outlined in this book.

Take-Aways

- VO2max plays a role in race performance, and we can develop it through appropriate training.

- The Billat 30-30 interval workout offers the promise of improving VO2max with less risk of overtraining. It comprises alternating segments done at maximum aerobic capacity speed with segments done at half power. These half power segments maintain the heart rate of the work phase while giving the working muscles a brief respite. Thus you can maintain a VO2max heart rate longer than if you performed a continuous max effort workout.

- We showed that the 30-30 workout protocol extends to paddling by accounting for the cubic relationship between paddling speed and applied power.

- The 30-30 interval recovery speed for paddlers should equal 80% of the work phase speed. You now know how to optimize your hard interval training in a paddling-specific way.

Further Reading

https://www.topendsports.com/testing/running-power.htm

F. García-Pinillos *et al.*, "Prediction of power output at different running velocities through the two-point method with the Stryd™ power meter," *Gait & Posture* **68**, pp. 238-243 (2019).

S. DeMarie, J.P. Koralsztein, and V.L. Billat, "Time limit at VO2max during a continuous and intermittent run," *J. Sports Med. Phys. Fitness* **40**, pp. 96-102 (2000).

V.L. Billat *et al.*, "Interval training at VO2max: effects on aerobic performance and over-training markers," *Medicine & Science in Sports and Exercise* **31**(1), pp. 156-163 (1997).

V.L. Billat *et al.*, "Intermittent runs at the velocity associated with maximal oxygen uptake enables subjects to remain at maximal oxygen uptake for a longer time than intense but submaximal runs," *Eur J Appl Physiol* **81**, pp. 188-196 (2000).

Appendix: Reading Mathematics

Introduction

As the late physicist Stephen Hawking noted about his book *A Brief History of Time*, "Someone told me that each equation I included in the book would halve the sales." Yet his book sold an astounding 25 million copies from 1988 through 2007. I'm not Stephen Hawking. But it appears that there is enough value in *The Science of Paddling* for readers to wade through the details willingly in search of useful nuggets. And I hope for the pleasure of learning new things.

The bulk of this book is grounded in physics and applied mechanics. And the language of physics and engineering is mathematics. In writing this book, I wanted to know more – to get at the *why* – and I believe readers want to know more as well. I could have dropped the rigor and asked you to believe my conclusions without supporting evidence. But I respect the work more than that, and I respect you, Dear Reader, more than that as well. So, I'm left with the convenient shorthand of science to show the basis for why: mathematics.

Mathematics need not be a mystery, a series of strange symbols and curlicues whose intent is to confound. The cure for that assumption is simply to think of mathematics as another language, a language you might not be very familiar with. Like all languages, math has rules, dialects, and idioms that spring from varied sub-fields of practice and application. Learning just enough of its grammar, words, and phrases will help you navigate the landscape and facilitate a more immersive experience as you read this book. You'll find that you don't have to *do* mathematics to *read* mathematics. There's a difference.

As an example, consider the language of equations. An equation expresses a concept. That's all there is to it. But equations often do not lend themselves to linear thinking like a computer program. An equation is more like someone describing an experience and the conclusions drawn from it, albeit in a concise and detailed way. For example, from Chapter 6 the equation

$$F_p = \frac{1}{2}\rho C_{d0} A \cos\left(\theta - \phi\right)\left(v_{hull} - v_{paddle}\right)^2 \tag{A.1}$$

is another way of saying, "I pushed off with my paddle, and the hull sped up." It synopsizes the relationship between propulsive force F_p and velocities v_{hull} and v_{paddle}. The other terms represent details via constant and variable definitions. Reading this equation also benefits from skill in looking at *limiting cases*; more on that below.

There are a few rules that will help you read and understand an equation's words and grammar. That's the purpose of this appendix. Think of this as a handy phrasebook for navigating the chapters in this book.

To realize this, we'll hew to an adage from my 9th-grade Algebra II teacher, Mrs. Verna Hazen. Mrs. Hazen's advice kept my classmates and me from getting bogged down in the "what *is* this?" of mathematics. Like my 9th grade self, we're usually more interested in what math *does*. I'll never forget her standing in front of the classroom. She would pause to consider her chalk-covered fingers, then look up and say, "Algebra is a game. And like all games, it has rules. If you follow these rules, you will *not* have any difficulty with algebra."

So, let's meet this challenge head-on, and review the rules – the words, grammar, and syntax – used in this book's mathematics, along with examples. Again, the goal isn't to have you *do* math but to read and get more from it. To achieve this, the material in the balance of the appendex is organized as follows:

- *Constants*
- *Variables*
- *Parameters*
- *Symbols*
- *Multiplication*
- *Vectors and Scalars*
- *Fractions*
- *Powers and Roots*
- *Functions*
- *Plots*
- *Derivatives*
- *Summations*
- *Integrals*
- *Equations*

We'll consider examples from selected chapters. A few topics include supplementary material if you'd welcome a bit more depth and historical context. The topics are arranged progressively, and the concepts build one upon the other. But you can read them in any order. Just remember that the art of reading mathematics requires some preliminaries. Your patience will be rewarded.

Constants

We deal with constants all the time without giving them a second thought.

For example, we use numbers to count things, and the meaning of numbers – i.e., their value – does not change. It would be disconcerting to discover that there are suddenly 37 minutes in an hour because 60 and 37 exchanged places in our numbering scheme. The more philosophical among you might contend that '37' and '60' are symbols that represent numeric quantities. But we also share a social pact that specific numeric symbols have particular and unchanging meanings. Hence, they are called *constants*.

The same agreement holds for quantities of measure like the meter (with symbol '*m*'), the second (with symbol '*s*'), or the liter (with symbol '*l*'). While typically employed to express measures, they too implicitly represent constants.

Geometry provides the constant representing the ratio of a circle's circumference to its diameter, π, with a value of 3.14159265358.... Jacob Bernoulli[117] gave us the constant e that forms the base of the natural logarithms and the natural exponential function, with value 2.7182818...

Other constants have their origin in repeatable physical measurements. These include the acceleration of gravity on the Earth's surface, represented by the symbol g; the density of water, typically represented by the Greek symbol ρ (pronounced "rho"); and the kinematic viscosity of water, ν (pronounced "nu"). Curiously, these constants depend upon specific environments. The gravitational constant changes as we move away from Earth's gravity well (thanks, Einstein!); the density and viscosity of water depend upon temperature. The only place where it is appropriate to equate g with 32.2 meters per second squared is Earth's surface. The values of water's constituent properties are usually specified at 20 degrees Celsius (68 degrees Fahrenheit) and must be adjusted otherwise. Still, we're all paddling on the surface of the Earth, and the temperature of a lake or river or the ocean usually doesn't change very quickly, so it's reasonable to refer to these as constants.

Variables

The number '6' as written or spoken is a concept. It is a convenient shorthand to describe something we see or think about, like the number of bottles in a six-pack or the number of letters in the English word 'friend.' We associate '6' with a specific quantity or value. But what if I wanted to represent the number of full spice jars in my pantry over time? Or the volume of gasoline in my car's tank while I drive? Or the number of letters in each of the words in the Czech language? Each of these describes a range of numbers. In those cases, we use a *variable*.

A variable is one level of abstraction higher than writing a number like '6'. We generally represent variables using a letter, such as x or t. Conventionally the symbols x, y, and z represent a spatial location or dimension. Since there are an infinite number of spatial locations, it makes sense to represent location compactly with a variable rather than via a very, very (very) long table of constant values.

The same goes for time, which can take on a range of values. By convention, time t is expressed relative to some fixed starting time or event. For convenience, this starting time is often (but not always) set to 0. We may define some starting event, like the initiation of an applied force or the beginning of a particular phase of movement, as the $t = 0$ reference. For example, this is how we do race timing.

While variables are, well, variable, they may nonetheless take on particular *values*. These

---

117 No, not that Bernoulli.

include the average speed of a river at a particular time and place or the instantaneour force of a cross-forward whitewater paddle stroke in the middle of the power phase.

With functions and equations, variables are either independent or dependent. More on this later.

Parameters

Parameters are variables that are set to a particular constant value. Then an equation is solved, or a simulation is run to determine the parameter's influence on results. For example, Chapter 6 studied a paddle's bend angle parametrically to determine its effect upon the synchronization of blade face angle, paddle force, and relative velocity.

With functions and equations, parameters are always independent variables; they don't depend on other variables. Parameters are usually not represented using special symbols; they look like other variables. Separating parameters and variables may sound like a distinction without a difference. The difference is in how we use the parameter. This distinction affords precision in our writing.

Symbols

One common stumbling block to reading mathematics is understanding Greek symbols. The simplest way to move past this hurdle is to realize that the Greek alphabet is used in mathematics like the letters used in the English language are part of a Latin alphabet. They're just letters. There's no need to overthink it.

Scientific papers were written in Greek during the Renaissance, much like Sanskrit served as the academic language of ancient India. There is a historical precedent for using Greek symbology in technical writing.

Here are some common Greek letters used in this book, along with their meaning:

ρ ("rho") – water density, a constant.

ν ("nu") – water's kinematic viscosity, a constant.

π ("pi") – the ratio of a circle's circumference to its diameter, a constant.

θ ("theta") – commonly used as a variable to represent angles.

ϕ ("phi") – commonly used to represent an angular offset, such as a paddle bend angle. Electrical engineers often use it to represent a phase lead or lag.

ω ("omega") – commonly used to represent "angular frequency." Angular frequency is two times π times the frequency in Hertz (aka the "circular frequency"). Angular frequency pops up because of geometry and trigonometry, which measure angles in radians rather than degrees. There are 360 degrees of arc in a circle, which equals one cycle (e.g., one time around the circle). There are two times π radians of arc in a circle. If we go around a circle once per second, the circular frequency is 1 Hertz[118] ('*Hz*') or one cycle per second, which is equivalent to subtending two times π radians of arc per second in angular frequency. It is often preferable to represent frequency using ω since this leads to more compact mathematical expressions. One would otherwise have to write $2\pi f$.

δ ("delta," lower case) – commonly used to express an infinitesimal change in some quantity.

Δ ("delta," upper case) – used to express a change in some quantity, though not necessarily an infinitesimal change.

∂ ("del") – another way to express an infinitesimal change in a quantity. Often found hanging around partial differential equations.

μ ("mu") – typically used as a prefix in units of measure, such as a micro-meter ('μm'), to denote one millionth.

τ ("tau") – commonly used as a time variable or to represent a lead or lag time.

Another confounding doodad in mathematical notation is the *subscript*. Subscripts are helpful labels that help related constants or variables be more easily grouped and identified. For example, a hull's dimensional drag coefficient is written as C_D, while the drag coefficient for a paddle blade when oriented perpendicular to the water's surface (or to an incoming flow) is C_{d0}. The '0' indicates an angle of 0 degrees to the flow direction (aka, "normal" to the flow). Subscripts sometimes indicate that a quantity is an average value, for example, expressing an average velocity as u_{avg} rather than using the more typographically cumbersome overbar notation.

Subscripts can denote sequences of variables. For example, a sequence of N variables S_1 S_2 $S_3 \ldots S_N$ can be written as S_i for $i = 1 \ldots N$. This signifies that the index 'i' ranges from 1 to N.

Other symbols represent common mathematical operations and relationships. These include many familiar as well as some unfamiliar examples:

$+$: addition

$-$: subtraction; also used to denote negative numbers (numbers less than zero)

$\times$: multiplication; see below for variants

$\div$: division; see below for variants

$=$: equal to

---

118 The unit of measure is named after the German physicist Heinrich Hertz, who played a key role in the early development of electromagnetic theory.

$\approx$: approximately equal to

$\neq$: not equal to

$<$: less than; the quantity on the left side of the '$<$' symbol is less than the quantity on its right

$\leq$: less than or equal to

$\ll$: much less than

$>$: greater than; the quantity on the left side of the '$>$' symbol is greater than the quantity on its right

$\geq$: greater than or equal to

$\gg$: much greater than

$\pm$: plus or minus. This symbol denotes both positive and negative values of what immediately follows it, or a range, such as for an error band

$\rightarrow$: goes to, meaning the quantity on the left goes to (or approaches) the quantity on the right. Often used to describe limits or limiting cases.

$'$: an apostrophe, usually referred to as "prime." Generally, this signifies a variant version of some quantity or function. It is also used in many applied math texts to denote differentiation with respect to an independent variable. We use the former here, not the latter.

∞ : infinity, the number which has no number larger than itself

(), [], {} : parentheses, brackets, and braces are used to enclose collections of operations, functions, constants, and variables. Sometimes this is done because of mathematical necessity when the items inside need to be evaluated separately. They also enable convenient groupings of terms with physical significance, like the difference between two velocities. I learned that you used parentheses first. You used square brackets if you needed another "enclosure" to group things already contained in parentheses. And if you needed to enclose things inside both, you used curly braces. These days many apps for formatting math equations just use larger and larger parentheses.

More specialized symbols for operators, and their uses, are defined below.

Multiplication

Why can't we settle on just one way of expressing multiplication? I have no idea. Sometimes multiplication is written in the familiar way, such as $2 \times \pi$ to denote two times pi. Alternatively, symbols are placed adjacent to signify multiplication, such as 2π. And finally, a dot ('•') can also represent multiplication via $2•\pi$. Wouldn't it be nice to settle on just one way of writing this? Yes, it would.

Vectors and Scalars

Any variable with directionality, or represented using more than one dimension, is a *vector*[119] Vectors have both magnitude and direction. Everything else is a *scalar*, which has magnitude but no direction. For example, velocity is a vector, and speed is its magnitude. The two are sometimes conflated. Constants like gravity's acceleration and water's density are common examples of scalars. And not to confuse things, but when all variables represent quantities that move in only one direction, the variables are often treated as scalars even though there is a directionality. Here, the directionality is implicit. This often leads to conflating speed with velocity in 1D. Distinctions like this may sound like "How many angels can dance on the head of a pin?" But for the sake of precision we must recognize them.

Many people think of vectors like arrows. An arrow has a length and points in a particular direction. This is a reasonable but imperfect analogy. A vector's length corresponds to its *magnitude*; magnitude is a scalar. The vector variable's dimensions along each coordinate system axis are its *components*. This is illustrated in Fig. A.1 for a force vector **F**. Note that vector variables are generally written in boldface while their components are not. In 2D, the projection of the vector's length on the x- and y-coordinate axes are its components F_x and F_y. In 3D, we further project the vector onto the z-coordinate axis to infer F_z.

By convention, a fluid velocity vector **V** has components in the x, y, and z directions often written as u, v, and w. So, if you see those, now you know.

As an aside, the ordering of the coordinate system's components for a 3D vector is crucial if you want the results you derive from it to be correct. Take your right hand, extend its fingers, and imagine that your wrist is at the origin of the 3D coordinate axes (where the coordinate axes intersect). Turn your hand so that the thumb points upward, and your palm and fingers align with the x-axis. Now rotate your hand about the wrist counterclockwise 90 degrees. Your fingers now point along the y-axis, and your thumb points upward along the z-axis in its positive direction. This is the so-called *right-hand rule*. Why is this important? I had a summer job where my project's lead engineer defined a model skyscraper's coordinate system with the z-axis pointing the wrong way: toward the ground. We measured structural and façade loads on scale models in a wind tunnel. If we hadn't corrected this axis definition, it might have affected the building's structural design. Math does indeed count, folks.

119 Constants can be vectors, such as the unit vectors.

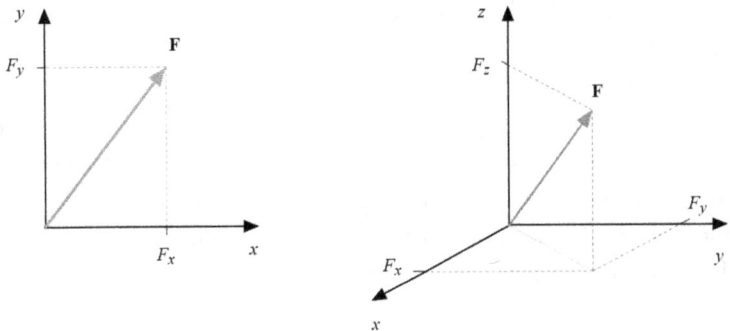

*Figure A.1: A vector **F** and its components in 2D and 3D.*

Fractions

I first encountered fractions in grammar school; I expect most of you had a similar experi-ence. Fractions are a way of representing division without dividing. Sometimes this is for convenience; other times, it's theater. But each of the following fractions serves a particular purpose:

$$\frac{1}{4} \qquad \frac{\rho}{2} \qquad \frac{1}{x}. \tag{A.2}$$

In the first example, you equate ¼ with dividing a whole (the '1') into four parts. Simple. In the second example, the symbol ρ, likely the density of water, is left "whole" in the numerator. When reading this fraction, plus any other mathematical terms adjoining it in a function or equation, the reader can associate mass per unit volume with the symbol ρ. This conveys more meaning than looking up the density of water in a table, dividing by two, and writing "0.49901 *g/cm³* @ 20C. "

We have our initial introduction to reading mathematics in the third fraction above. One superpower granted to those who learn to read mathematics is understanding the implications of what is written. Let's assume 'x' is a variable in the third fraction's denominator above. What are the implications for different values of x?

You might ask, "What do you mean by implications? It's just a fraction!" Sure, it's a fraction. But what value(s) does the fraction take as the variable x changes? When x equals 1, the fraction equals 1 – check. When x becomes smaller than 1, the fraction becomes greater than 1. This is because dividing a number (here, the numerator 1) by a smaller number (here, the denominator) produces a result greater than 1. As x becomes smaller and smaller, the fraction becomes larger and larger; try a few values for x and see. As x approaches 0, the fraction becomes enormous; in the limit, it becomes infinite. If x becomes larger than 1, the fraction becomes less than 1. This is because dividing a number (here, the numerator 1) by a larger number leads to a result that is less than 1. As x becomes larger and larger, the fraction becomes smaller and smaller; try a few values for x and see. In the limit, as x approaches infinity, the fraction goes to zero. Finally, if x is a negative number, the fraction is also negative. This is because $1/x$ is an odd rather than an even function, something we'll learn more about below.

This exercise is an example of *reading* mathematics. Rather than glossing over the fraction $1/x$, we spent a little time exploring its implications in light of the variable denominator. You can similarly investigate fractions with variable numerators, or where both numerator and denominator are variables. Mastering just this one *active reading* skill will take you very far in extracting meaning from what might otherwise be a jumble of symbols and squiggles on the page. Don't underestimate its power to inform.

Powers and Roots

One bit of notation you'll see sprinkled throughout this book is a constant or variable with a numeric superscript, like x^2 or v^3, or even $L^{2/3}$. These superscripts (generally) represent a power of the constant or variable; you'll sometimes encounter the term "raised to the power of…" to signify this operation. Two common powers have names: x^2 means the variable x is "squared" or multiplied by itself; v^3 means v is "cubed" or multiplied three times. The superscript is called the *exponent*.

Ever wonder where the terms "squared" and "cubed" came from? If you square a number, the result equals the area of a square with sides equal to the number. If you cube it, the result equals the volume of a cube with sides equal to the number. These are illustrated in Fig. A.2 for the square of 2 (area = 4, which you can prove for yourself by counting) and the cube of 3 (volume = 27, which you can also prove for yourself by counting).

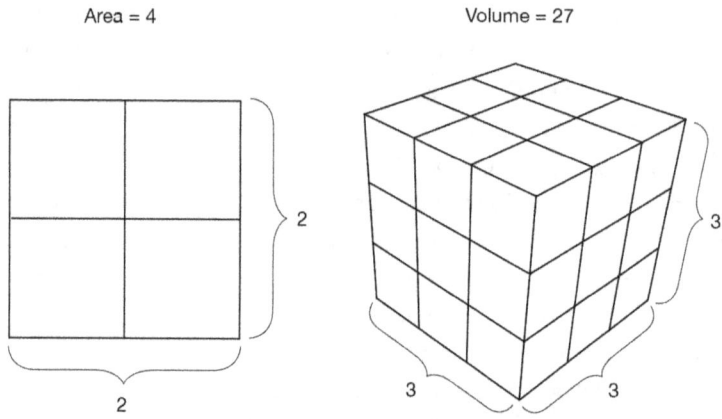

Figure A.2: Square and cube, illustrated.

Roots are the *inverse* operation of powers. Referring to the figure, you can see that the square root of 4 equals 2; the square root is the number that would be "squared" to construct a square of area equal to 4. Similarly, the cube root of 27 equals 3; the cube root is the number that would be "cubed" to construct a cube with a volume equal to 27. It gets a bit awkward after cube roots (and cubes, for that matter) since a geometric construction like those illustrated in the figure would have to be drawn in a 4-dimensional space or higher. Don't worry; I don't know how to do that either.

Fractional powers can be confounding since they aren't as intuitive as whole powers. Fortunately, there are a few simple ones that will get you started. For example, we can write the square root as raising something to the power of one half:

$$\sqrt{c} = c^{\frac{1}{2}}, \tag{A.3}$$

and the cube root is the same as raising something to the power of one third:

$$\sqrt[3]{P} = P^{\frac{1}{3}}. \tag{A.4}$$

Next, a constant or variable raised to a negative power is the same as dividing 1 by the constant or variable raised to the positive value of that power. This is most easily understood with a couple of examples:

$$x^{-2} = \frac{1}{x^2} \qquad P^{-\frac{1}{3}} = \frac{1}{P^{\frac{1}{3}}} = \frac{1}{\sqrt[3]{P}}. \tag{A.5}$$

Functions

When I was a high school senior, I took a Calculus I class at a nearby college since my school didn't offer this subject. I had to make a conceptual leap to embrace functions. Before that class, I had always done relatively concrete mathematics: add some numbers, factor an equation, solve a quadratic, or plot a line from a table of values. There was almost nothing abstract about it, and we seldom looked to generalize our work.

But I learned functions are no big deal and a very useful concept. Think of a function as an input-output black box, as suggested in Fig. A.3. You give it an input, and an output emerges. "Function" is just another way of saying "one or more mathematical operations performed in a prescribed way upon one or more constants, variables, or functions." The last part about functions performing operations on functions is a bit abstract – just hold tight.

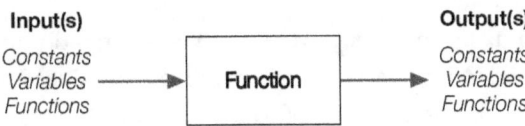

Figure A.3: A function as an input-output black box.

A function may comprise familiar arithmetic operations like addition, subtraction, multiplication, division, powers, and roots. Trigonometric functions like sine and cosine are functions: give them an input angle and they return an output value. They can be used to construct larger, more complex functions.

We can write a function that operates on a single variable x using shorthand such as $f(x)$, pronounced "f of x." 'f' is the function as well as its name. Any other letter or symbol can designate it; f is just one choice and nothing more. The input variable x, often referred to as the function's *argument*, is an *independent variable* since the function depends upon it and not the other way around. If the output of the function is another variable, say y rather than a constant value, then y is called a *dependent variable* since it depends upon the function f as well as the independent variable x.

We can express this succinctly, and somewhat abstractly, as

$$y = f(x). \tag{A.6}$$

Note that we haven't specified the function f or what it does. This is by design; it was here that I tripped up so often when first introduced to this terminology. But the power of this very general form of representation lets us define f as, for example,

$$f(x) = \frac{1}{2} \rho A C_{d0} \cos(x), \tag{A.7}$$

or

$$f(t) = m\frac{dV}{dt} + C_1 V + C_2 V^2, \tag{A.8}$$

which is a governing equation for a system with both linear and quadratic drag. After defining the function we don't have to write down all the operations that comprise it again; the reader can refer back to the details as needed. Expressing functions this way is a convenience, a shorthand. It's like naming the thing you are sitting on "chair" without listing each of its parts, the materials from which it is constructed, its dimensions, color(s), and upholstery fabric and pattern every time you refer to it. It's more compact, too.

A function can be as simple as one arithmetic operation, such as addition. Or we could define a function – we'll call it the "Douglas Adams" function[120] – that multiples an input variable x by 42:

$$f(x) = 42x. \tag{A.9}$$

This means you've been using functions ever since you started doing multiplication tables. Surprise! Now this function isn't very interesting. But it provides another straightforward example in reading mathematics. If x goes to zero in our Douglas Adams function f, we note that f goes to zero as well. If x becomes very large, f also becomes large. Forty=two times as large as a matter of fact. If x is negative, f is also negative. This is how you actively read a function. You look for where it equals zero, what happens if its input variable(s) go to zero or become large, how it behaves with positive versus negative inputs, etc.

Consider now the function

$$f(C) = 1.8C + 32. \tag{A.10}$$

---

120 Any *Hitchhiker's Guide* fans out there?

If $C = 0$ then f returns 32; if $C = 100$, f becomes 212; if $C = -17.8$ then f goes to (approximately) 0. We have defined a function f that converts from degrees Celsius to degrees Fahrenheit. As you can see, functions are no big deal, and nothing to get wound around the axle about.

Things get more interesting with

$$f(x) = \frac{x^2}{(1-x)^3}.$$ (A.11)

When $x = 0$, this function equals zero – try it and see. What happens when the independent variable x gets large? By "large," we mean that x assumes a magnitude, positive or negative, large enough that the 1 in the equation's denominator can be ignored in comparison. In that case, the function f is approximated by

$$f(x) \approx \frac{x^2}{(-x)^3} = \frac{-1}{x}.$$ (A.12)

As x gets larger and larger in magnitude, f will become smaller and smaller and eventually approach zero. The negative one in the numerator indicates that the value of f will be negative for large positive values of x and positive for large negative values of x. Remember, this approximation is only valid for values of x with magnitude much larger than 1. And finally, something interesting happens when $x = 1$. The numerator then equals 1. The denominator, however, equals 0. Dividing anything finite by 0 equals infinity. Therefore f becomes infinite at $x = 1$. Who knew there were so many things going on in this simple equation? You do, because you are reading mathematics.

There are a few common trigonometric functions worth knowing about, like sine and cosine, along with the transcendental functions of exponentiation and logarithms. We'll save those for the next section on plots. There are also two key functional operations from calculus: derivatives and integrals. We'll visit with them later as well.

Plots

Plots are visual representations of the mathematical relationship between variables. They are also a means for visual learners to read mathematics, complementing how we read functions as described above. An example appears in Fig. A.4.

This plot compares x and x^2 over a range of x values. While it doesn't look like much, this plot provides some helpful tips for reading mathematics. First, notice what happens when $x = 1$: both curves intersect. Between 0 and 1, the magnitude of x squared is smaller than the magnitude of x. This is true for numbers between 0 and 1 for any power of x – higher powers of x are always smaller in magnitude than lower powers of x when x is between 0 and 1. And above this, the magnitude of x squared is greater than x. This is true for all whole number powers of x – higher powers of x are always larger than the lower power when x is greater than 1. These two "small" observations come in handy repeatedly in applied mathematics when we look at limiting cases. For example, when is one variable larger than another? Can we simplify an expression by dropping terms that are much smaller than others? Master these straight-forward observations, and you will go far.

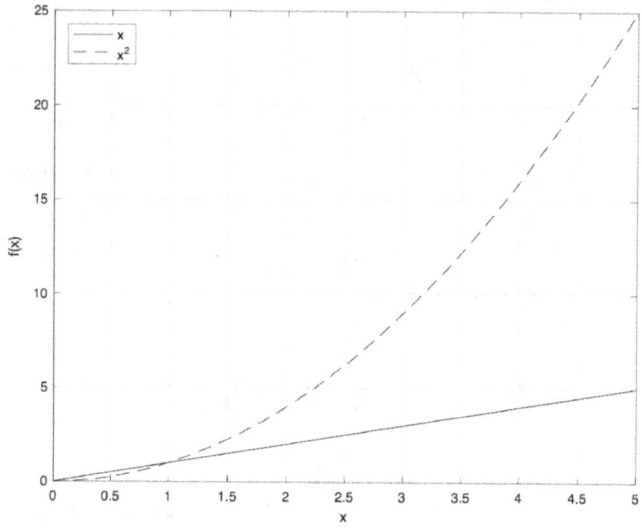

Figure A.4: Plots of x and x^2 vs. x.

Where do plots come from? The 2D plot in Fig. A.4 above is a way of representing the *relationship between quantities* in a Cartesian[121] coordinate system. These often correspond to physical quantities like space, time, force, or power. In that way, Fig. A.4 is like a map, with two coordinate axes – here, values of the independent variable x (with horizontal coordinate axis called the *abscissa*) and the values of two functions of x (with vertical coordinate axis called the *ordinate*), rather than latitude and longitude.

When I first learned to plot functions, I drew a Cartesian grid, created a table with ordered pairs of x values and its dependent function(s), then placed a dot on the grid for each pair. Nowadays, I use programs like MATLAB or Excel to generate these plots. Or, because I've plotted many functions over the years, I sometimes can draw a good approximation in my notebook to get started.

When reading any plot, first look at the abscissa and ordinate ranges, as well as their labels. Is this a plot of how something varies over time? Over space? With respect to a variable parameter? This contextualizes the plot and informs its interpretation. Then, how many relationships – e.g., separate curves – are presented? If there is more than one curve, what does each represent, and how do the curves relate or differ? Is something getting small, or large, or changing from positive to negative for one or more values or ranges of the independent variable? Here you're looking for trends since the plot represents the

---

121 The Cartesian coordinate system is named after French philosopher and mathematician René Descartes. His visual way of representing the relationships between numbers bridged the gap between algebra and geometry.

relationship between things in a visually accessible way. This is plotting's greatest value. In short, you're asking the plot, "What are you telling me?"

Consider the plots of x and its square and cube roots in Fig. A.5. Fig. A.5, like Fig. A.4, shows the relationship between x and two of its powers. These powers are fractional: the ½ power (i.e., the *square root*) and the ⅓ power (i.e., the *cube root*). Like in Fig. A.4, something interesting happens with these three functions around $x = 1$. Using your ability to read mathematics, what can you say about larger vs. smaller fractional powers on either side of $x = 1$?

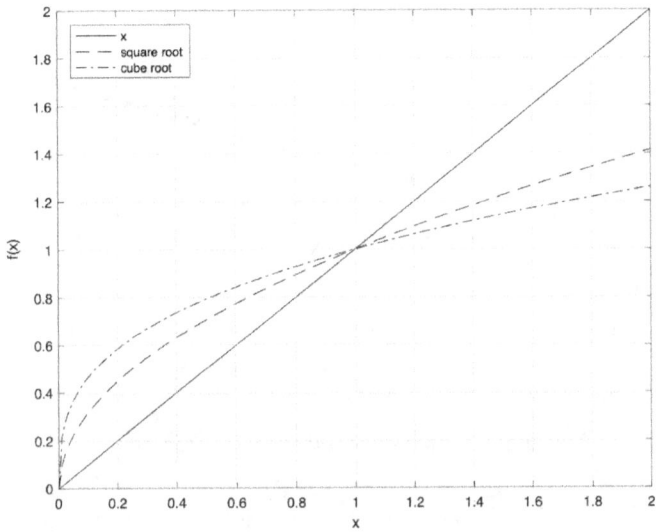

Figure A.5: Plot of x and its square and cube roots vs. x.

You'll sometimes find that a plot's horizontal or vertical axis scales look weird. They may not be evenly spaced. Suppose the independent variable or corresponding functional values extend over a wide range of values. In that case, it sometimes makes sense to use a logarithmic spacing for one of the plot axes.

Logarithmic spacing isn't all that unfamiliar. That's how your hearing perceives sound, both in how you can pick out its various tones as well as how you perceive sound volume changes. For instance, the keys on a piano are assigned to notes such that every octave, or frequency doubling, corresponds to twelve keys. Using an equal number of keys for successively wider frequency bands is an example of logarithmic (aka, "log") spacing. Or, you may have seen sound levels expressed in decibels (aka, "dB," a unit of measure named after Alexander Graham Bell). Decibels are logarithmic sound amplitude values since our hearing more easily distinguishes successive amplitude doublings. Our hearing range covers 120 dB of amplitude from softest to loudest, or an amplitude span of 1,000,000,000,000 (one trillion). The graph would be unreasonably large if you plot numbers from zero to one trillion on a linear scale.

Fig. A.6 is an example plot with log axes. Here, we revisit a few old friends: powers of an independent variable x. Since we (now) know that taking a variable's root is the same as raising it to a fractional power, Fig. A.6 tells us something interesting. When plotted on log axes, *any* power of a variable is simply a straight line. This doesn't mean that powers of a variable are straight lines, only that their plots on *logarithmic axes* are. The various

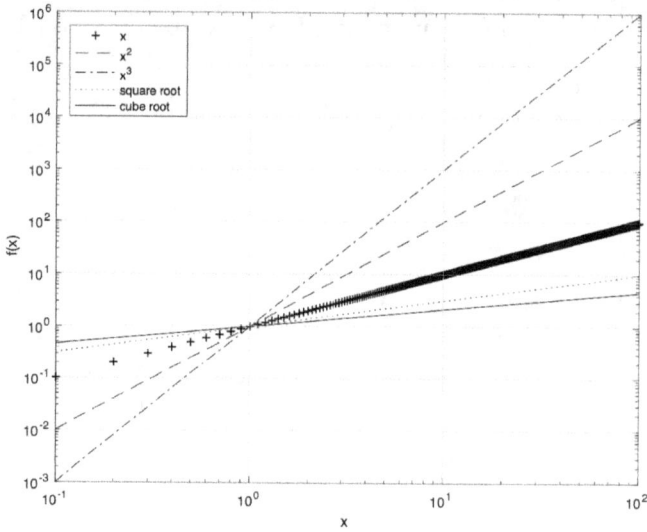

Figure A.6: Logarithmic plot of various powers of x.

slopes of the straight lines in Fig. A.6 simply represent the corresponding power of x, aka its *exponent*. Fig. A.6 embodies the following mathematical identity:[122]

$$\log x^n = n \log x. \tag{A.13}$$

Like our logarithmic plot, this identity shows that the log of anything raised to a power (whole or fractional) is the same as the power times the log of the said thing. For those who grew up using "slope and y-intercept" rules for plotting straight lines, Fig. A.6 uses $\log x$ as the independent variable and the exponent as the slope. This insight into the logarithm of powers led me to write Chapter 18.

We can use plots to provide visual summaries of a few common trigonometric and other transcendental functions. First, Fig. A.7 is a plot of the sine and cosine functions. Note how the sine and cosine are identical except that they are shifted along the plot's t-axis by $\pi/2$. Their amplitudes vary between -1 and +1. And the sine function is zero when its argument t is zero, while cosine equals one there. This pattern repeats itself along the horizontal axis in 2π increments. If the horizontal axis in Fig. A.7 corresponds to time, the spacing corresponding to this repetition is called the *period*. If the horizontal axis in Fig. A.7 corresponds to space, this repetition distance is the *wavelength*.

---

122 Math identities are truisms that show two things that are always identical. Hence, "identity."

Fig. A.8 plots the exponential and natural logarithm functions – note that exp(x) is the same as e^x. Aside from noting where each function grows or decreases in magnitude and becomes negative, you see that the natural logarithm (i.e., the logarithm in base 'e' – more on that in a moment) of 1 equals 0. In contrast, the exponential of 0 equals 1.[123]

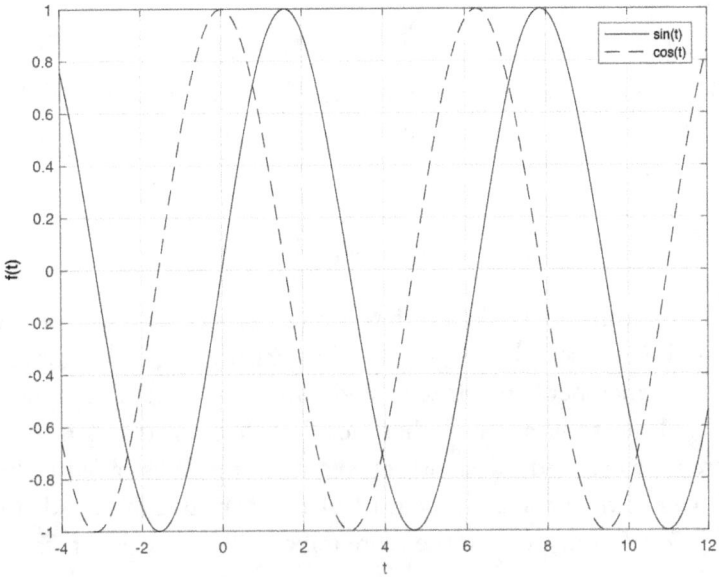

Figure A.7: Plot of sine and cosine.

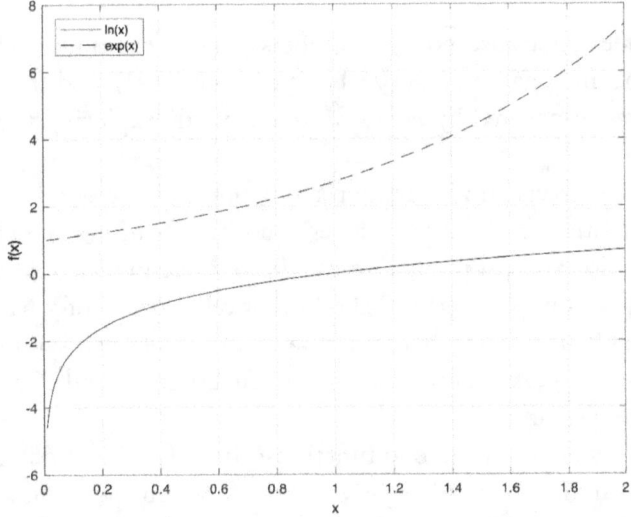

Figure A.8: Plots of natural log and exponential.

123 Note that raising anything to the 0-th power equals 1, including negative numbers. In applied mathematics, raising zero to the zeroth power also equals one. Other branches of math may leave it undefined.

This curious symmetry arises in part from the relationship between the exponential and the natural logarithm: they are inverse functions of each other. In other words, the natural logarithm of the exponential of an independent variable returns the independent variable. This is because the inverse function of a function returns the argument; ln(x) and exp(x) are examples. You'll see this simply via the equation

$$\log_e \left(e^x \right) = x .$$
(A.14)

The exponential function merely raises the number e to the power of x. We (now) know that taking the log of anything raised to a power is equal to that power times the log of said thing. Which means that

$$\log_e \left(e \right) = 1 .$$
(A.15)

So, the natural logarithm can be defined based on its properties. It is the inverse function of the exponential (hence the "base e"). And the logarithm of powers does that nifty linearization thing. Remember Mrs. Hazen's advice from the start of the appendix, and don't worry about what logarithms are, but what they do. Don't overthink it.

For extra credit, where did this constant e come from? 'e' is called Euler's Number in honor of the Swiss mathematician Leonard Euler – even though he didn't discover it. It is the basis of *Euler's Identity*, one of the more remarkable and elegant statements in all of mathematics:

$$e^{i\pi} = -1 .$$
(A.16)

This identity includes the transcendental numbers e and π, the transcendental function of exponentiation, the imaginary number i (the square root of -1), and the negative number, -1. But e itself was discovered by Jacob Bernoulli while exploring the mathematics of compound interest.

Lastly, plots help us visualize whether a function has *even* or *odd symmetry*. *Even* functions return positive results for all values – positive or negative – of its independent variable. *Odd* functions return positive results for positive inputs and negative values for negative inputs. For example, Fig. A.9 shows that odd powers of x have odd symmetry about $x = 0$. Fig. A.9 also shows that even powers of x have even symmetry about $x = 0$. Along the same lines, Fig. A.7 shows that the sine function has odd symmetry about $t = 0$, while cosine has even symmetry.

Why do we care about even and odd functions? It is often helpful to understand when or if a function flips from positive to negative. If the function represents a physical process, this may indicate (for example) a change of direction, acceleration versus deceleration, or a change in the direction of power flow from source to sink. The product of two even functions is even. The product of two odd functions is even. And the product of an even and an odd function is odd. Factoring a function into a product of even- and odd-symmetric parts lends insight into the underlying processes and their interrelation.

As you have seen, plots make several mathematical concepts more accessible and concrete. Use them liberally and wisely.

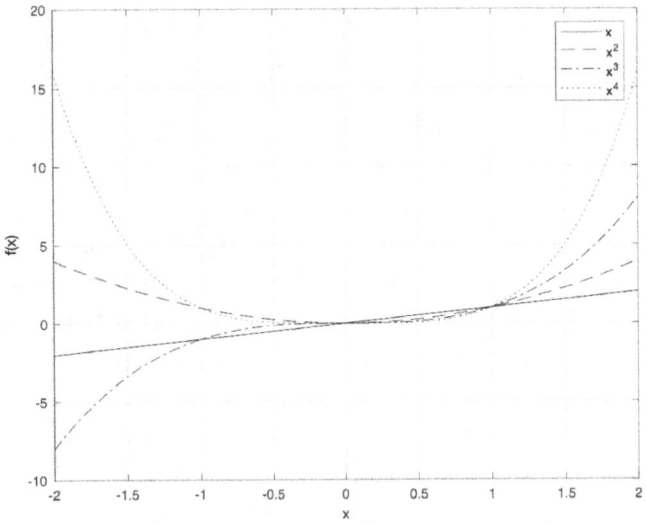

Figure A.9: Powers of x showing their odd and even symmetries.

Derivatives

Ordinary derivatives[124] represent the rate of change of a function with respect to an independent variable. You might say that the derivative is performed "with respect to" that variable, as in the "derivative with respect to time" or the "derivative with respect to x." While derivatives are typically one's first entry point into calculus, we won't be asking you to compute them (unless you want to), only read them. It's not complicated; there are straightforward rules for computing them.

A few of the most common derivative expressions you'll encounter in this book define the relationship between position x, speed v, and acceleration a:

$$v = \frac{dx}{dt}, \qquad a = \frac{dv}{dt} = \frac{d^2x}{dt^2}. \tag{A.17}$$

What looks like fractions above actually represent derivatives. The notation comes to us from Gottfried Leibniz, who developed calculus at around the same time as Isaac Newton.[125] What looks like squared quantities mean taking a *second derivative*: acceleration is the time

---

124 There are also partial derivatives, which are derivatives with respect to one variable when the differentiated function depends on two or more independent variables. The symbol for a partial derivative is a bit different, but the rest is a matter of housekeeping.

125 Newton placed dots over variables to denote derivatives. You find this notation in many dynamics texts. Lagrange used apostrophes, a notation that you find in applied mathematics texts.

derivative of velocity, which is the time derivative of position. Consequently, acceleration is the derivative of the derivative of position, or the "second" derivative of position with respect to time. Note that the time dependence of position, velocity, and acceleration is implicit in the above expressions; this saves writing, for example, velocity as $v(t)$. Lastly, you may sometimes encounter the term "differentiation," which means to take the derivative of some function. It's just related terminology. And that's all you need to know; skip to the next section if you wish.

The balance of this section is a sidebar for those who want a more intuitive understanding of the derivative operation.

First, note that the 'd' in the derivative notation suggests an infinitesimal change in the adjacent variable. So dx represents an infinitesimal change in position x, while dt represents an infinitesimal change in time t.[126] And what is an "infinitesimal"? An infinitesimal is a quantity that is smaller than *any* number but not equal to 0. This definition gives rise to the phrase "arbitrarily small." How does this relate to derivatives representing a rate of change?

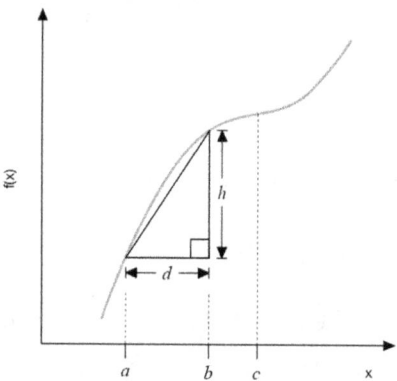

Figure A.10: A continuous function.

It's easiest to explain using a plot. Consider the arbitrary continuous[127] function $f(x)$ plotted in Fig. A.10. Where do you see this function changing more rapidly or less rapidly with respect to x? Certainly, at $x = a$, the function is changing more rapidly than at $x = c$. Small changes in the independent variable x around $x = a$ lead to greater changes in $f(x)$ than around $x = c$. How do we infer this? From the slope of the function $f(x)$ around those locations. The slope of a function reflects its *local rate of change*. If the slope equals zero over a range of the independent variable, there is no change in the function's value.

---

126 Infinitesimal quantities were considered heretical during the Middle Ages because they contradicted Aristotle's philosophy, which was widely accepted at the time. Aristotle contended that physical reality was continuous and could not be reduced to atoms or infinitesimals. And don't forget Zeno's Paradox. You can read about all this and more in *Infinitesimal: How a Dangerous Mathematical Theory Changed the World* by Amir Alexander.

127 For those of you keeping score at home, if the function has sudden jumps in amplitude (e.g., is not continuous), its derivative technically doesn't exist at those discontinuities. You must invent a whole new class of functions called *generalized functions* to address this limitation.

If the slope varies "a lot," then the function varies "a lot" as well. So, if $f(x)$ represents the position of your hull over time, and x represents time, then a steep slope in some portion of the position versus time plot means you're moving fast. A gradual slope means you're moving slowly, and a flat slope means that your position over time isn't changing and you're stationary. Let that sink in.

Now about those infinitesimals. Consider the right triangle shown in Fig. A.10, where two of its vertices intersect the function $f(x)$ at $x = a$ and $x = b$. The triangle's hypotenuse connects these two intersection points. The hypotenuse is thus coincident with the *secant* to $f(x)$ there. The side labeled 'd' is the triangle's width, while 'h' is its height. The ratio of this triangle's height to its width corresponds to the slope of the hypotenuse and is coincident with the secant; I recall hearing this ratio referred to as "rise over run." We can define the slope of the hypotenuse via values of the function $f(x)$ at the locations $x = a$ and $x = b$ (which equals $a + d$) and the width d:

$$\text{slope} = \frac{f(a+d) - f(a)}{d} . \tag{A.18}$$

For small values of the run d, this fraction's numerator equals the incremental change in a function (such as position). Then the denominator is the incremental change in its independent variable (such as time). This is how we interpret Leibniz's derivative notation.[128] As the triangle's width d gets smaller and smaller – as it becomes *infinitesimal* – the secant approaches the *tangent* to $f(x)$ at $x = $ a. This tangent is the function's slope at $x = a$ and therefore equals its derivative there. So, to determine a function's derivative at a location $x = a$, we use the equation above and let d go to zero. To derive a function's derivative everywhere, we can repeat this process for every value of its independent variable. Or, may choose to learn the derivatives of a few common functions, look up the derivative in a book or online, or both.

Summations

Summations are just a compact way of representing a sum of many parts. Here's an example:

$$\text{Total} = \sum_{i=1}^{N} s_i . \tag{A.19}$$

In the equation above, "Total" is the summation's total. The large sigma ("Σ") indicates that whatever sits to its right are the things summed: the *summand*. In the example above, several quantities represented by the symbol s are added together. The subscript 'i' is an index for the things summed. The subscript beneath the sigma indicates that we'll start the summation with the first index $i = 1$. This corresponds to the first item s_j; note that the sequence can start at an index other than 1. The sum may sometimes only be over a subset

---

128 The mathematical formula here looks like Newton's "difference quotient" – recall that Newton was a co-inventor of calculus.

of the items s_i, so read the summation's notation carefully. The superscript above the sigma indicates that the last item here will have index N, where N is a constant. In many cases, a specific number appears here, or sometimes even infinity.

The summation above says, "add together N items ranging from the first to the Nth." That's it. Summations are a convenient and compact way of writing

$$\text{Total} = s_1 + s_2 + s_3 + s_4 + s_5 + \ldots + s_N. \tag{A.20}$$

One example of a summation's utility is approximating the area under a curve. Let's say that someone owns a lakeshore lot. The shoreline has a shape described by the curve in Fig. A.11. Their lot is bounded on the west by $x = 1$ and on the east by $x = 5$. They want to know the lot size so as not to overpay their property taxes.

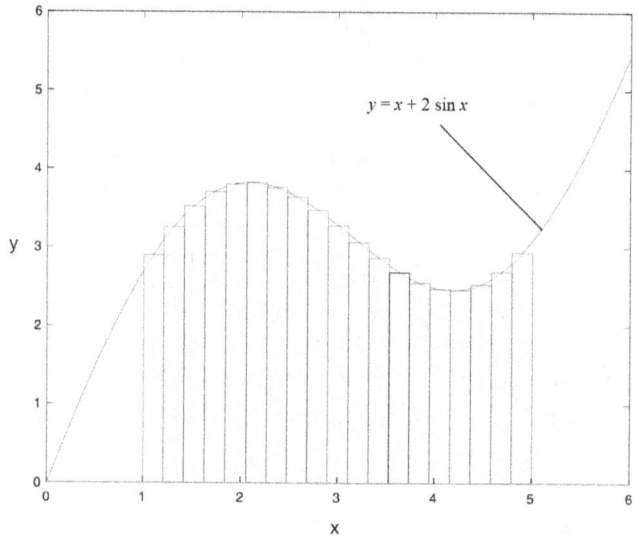

Figure A.11: Approximating the area under a curve.

Since this shoreline has a shape described by the function $x + 2 \sin x$ they can divide their lot into a series of plank-shaped rectangles with uniform width d. Then, compute the value of the shoreline shape function at the midpoint of each rectangle to get the corresponding rectangle's height. We multiply these two quantities to calculate the area of each rectangle. Sum these all up and voila! They'll obtain the plot's area.

If there are N rectangles, and each rectangle has area A_i, then the acreage is

$$\text{Acreage} = \sum_{i=1}^{N} A_i. \tag{A.21}$$

For extra credit, if we denote the value of the shoreline shape function at the center of each rectangular area by f_i, then each area is

$$A_i = d \times f_i,$$ (A.22)

where the width d is a constant.[129] Then our acreage summation can be written as

$$\text{Acreage} = \sum_{i=1}^{N} d \times f_i = d \sum_{i=1}^{N} f_i,$$ (A.23)

where we have taken d outside of the summation since it is constant – this is just multiplication's *distributive property*.

Now consider this last summation in light of Fig. A.11. Intuitively, we see that our acreage estimate will be more accurate by using narrower and narrower rectangles to approximate the area under the shoreline curve. We'll have to use more terms in our summation, e.g., N will become larger, and d will consequently become smaller. So, if we imagine the width d becomes infinitesimally small, and consequently N becomes larger and larger until it approaches infinity, we end up with Zeno's paradox.[130] Or we discover integrals.

Integrals

There are two types of integrals: definite and indefinite. Definite integrals assign a value to a function over a finite interval of its independent variable by adding together – i.e., integrating, or summing – an infinite number of infinitesimal areas. This is illustrated in Fig. A.12 for the function $x + 2\sin x$ from 1 to 5; the shaded region represents the area. This is the continuous version of the acreage problem illustrated in Fig. A.11. You may have already deduced that definite integrals and sums are related. The integral depicted in Fig. A.12 is represented symbolically by

$$\int_{1}^{5} \left(x + 2 \sin x \right) dx \cdot$$ (A.24)

The integral sign is the stretched 'S'. The subscript '1' is the integral's starting point, also referred to as the lower limit of integration. The superscript '5' is the operation's end point, referred to as the upper limit of integration. Everything to the right of the integral sign aside from the dx is the *integrand* or the thing to be integrated. 'dx' is a differential of the independent variable x, indicating that integration is performed with respect to this variable. It also reflects the definite integral's role in computing the area under the curve

---

129 It doesn't have to be constant but doing so makes things simpler.

130 Zeno couldn't conceptualize how an infinite number of infinitesimal things added together could yield a finite result. His famous paradox is about motion. The argument proceeds as follows: to move from where you are to some destination, you must move halfway there. You then move halfway from this new location to the destination. And so on, halving the distance with each iteration. You always move halfway from where you currently are to the destination. Zeno took this to mean that you would move an infinite number of times halfway but never reach the destination; you'd always end up halfway to it. Infinitesimals and calculus resolve this paradox.

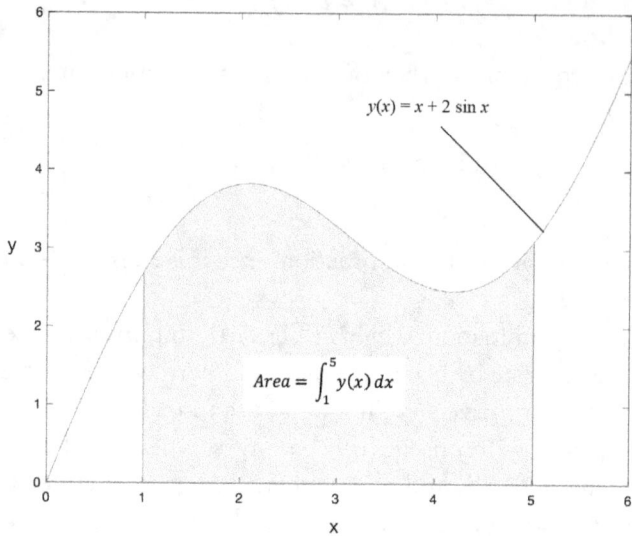

Figure A.12: Area under a curve.

defined by the integrand since dx is infinitesimal. Definite integrals compute areas; that's the most straightforward way to think of them.

More generally, we write the definite integral of a function of one variable $f(x)$ as

$$\int_a^b f(x)\,dx \tag{A.25}$$

where a and b are the integral limits and x is the independent variable. Whenever you see an integral written with limits like this, the result will be a constant since the integral "integrates away" the independent variable. The result might be expressed using other constants, but the independent variable will no longer appear after taking the integral. The result will be some amount, such as an area.

The thing about functions is that they can have positive or negative values over certain ranges of their arguments. Consider the function depicted in Fig. A.13: $2 - x + \sin(3x)$. Between 0 and 2.5, it takes on both positive and negative values. As a result, the definite integral of this function over the interval from 0 to 2.5 comprises two areas with a positive value and one "negative" area. We tend to think of the area as greater than zero. Referring to our lakeshore lot example, what is negative acreage?

But integration is literal. This means there can be negative areas as well as positive ones. For example, let the line $y = 0$ in Fig. A.12 describe the original shape of a sandy shoreline, and the function $2 - x + \sin(3x)$ describes the desired shoreline shape. Then the integral tells you how much sand to buy for building the new shoreline, given how much you can reuse by excavating the "negative area" portion. This made-up example underlies an important fact: definite integrals can return a zero value. And when they do, usually something important and worth your attention is going on.

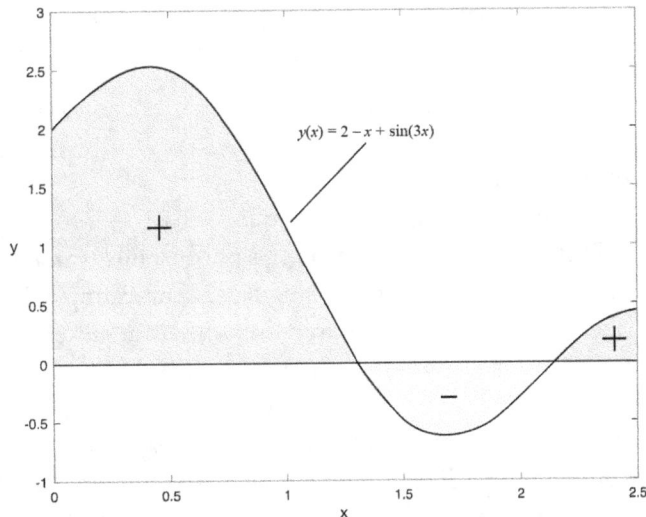

Figure A.13: Integration of a function having positive and negative areas under the curve.

You can perform integration over multiple dimensions if the integrand depends on more than one independent variable. For example, the flow velocity across a hull's wake varies over both the wake's width as well as its depth. To compute the force from the flow's dynamic pressure due to the wake, you need to integrate over the wake's cross-sectional area – see Chapter 2's Extra Credit section for an example. The result is still a constant; remember, definite integrals "integrate away" independent variables.

You might now be wondering what all the fuss is regarding the terminology "definite integral" rather than just "integral." It's because there is another type called an *indefinite* integral. They are easy to spot since they have no limits of integration. Here's an example:

$$h(x) = \int f(x)\,dx. \tag{A.26}$$

Notice that indefinite integrals are written (sometimes implicitly) with a function on the left-hand side of the integral sign. They don't "integrate away" the independent variable but instead return another function.

For extra credit: the functions h and f in the expression above have a unique relationship: h is the integral of f, while f is the derivative of h. This last sentence requires a few qualifications that aren't important here. But a nifty piece of mathematics called the *Fundamental Theorem of Calculus* proves that integration and differentiation are inverse operations; they are two sides of a coin. Therefore, you can move from position to velocity to acceleration via differentiation and from acceleration to velocity to position via integration. Useful stuff.

Equations

Equations are just various combinations of the things described above, where one thing is set equal to another, hence, "equation." That's it; you're done. Here's an example:

$$F_p = \frac{1}{2}\rho C_{d0} A \cos\left(\theta - \phi\right)\left(v_{hull} - v_{paddle}\right)^2 . \qquad\qquad (A.27)$$

So how do you put all this together to read an equation? Here are some guidelines:

First, take your time.

In case you forgot, take your time!

Look at both sides of the equals sign. What equals what?

Then look at the groupings of terms – the clumps of constants, variables, operators, etc. For each element, what does it mean in plain language? For example, is one grouping an integral, and what does it represent? Force; impulse; something else? In this step, you're just scanning the equation for its overall structure to determine what concept(s) is(are) being presented.

The equation may contain constants. In a well-written paper or text, these are defined by the author. Certain constants like π, g (the acceleration due to gravity), or e are well known in various fields., and the author may assume you are part of the in-crowd and familiar with them.

Next, identify the variables and determine their meaning and how they are used. The equation will define the relationships between these variables.

Does one side of the equation equal zero? If so, can you determine what causes everything on the other side of the equation to equal zero? Suppose the non-zero side comprised two more things added (or subtracted) together. In this case, some terms must cancel each other to yield zero. What does each part of this summation mean? And what does it mean if they cancel each other?

Does a variable appear by itself on one side of the equals sign? If so, that variable may be an output like a force or velocity.

Try to visualize how changing one variable affects the output / unknown quantity. Or search for limiting cases, such as when one variable gets large or small compared to another (or large/small in an absolute sense). What causes the numerator or denominator to equal zero if there is a fraction? Or what causes a function's argument to go to zero? Or what happens when a variable goes to zero? Keep in mind what range of values the variable(s) can take on. This may constrain one or more variables to physically meaningful ranges.

So, let's go back to the equation from the beginning of this appendix, presented in Equation (A.27), and read it. The author has defined F_p as the propulsive paddle force, A as the blade area, C_{d0} as the blade's drag coefficient at normal incidence, θ as the shaft angle with respect to the vertical, ϕ as the bend angle of the blade with respect to the shaft, and v_{hull} and v_{paddle} as the forward velocities of the hull and blade, respectively. The constant ρ is the density of water since this equation has something to do with paddling.

The propulsive force F_p stands by itself on the left-hand side of the equation. This means it is an output, e.g., the result of things on the other side of the equation. Since we know that the hull velocity, paddle velocity, and shaft angle vary over time, we infer that the blade force is a function of time.

The fraction and the water density are fixed constants. The blade drag coefficient is a constant as well.

The equation shows that the blade force varies in direct proportion to the blade area *if every other term in the equation stays the same*. Don't forget this qualifier since it will prove helpful in a moment. You could infer that this relationship means if you want boundless propulsive force – and who doesn't? – you just need an enormous paddle blade. This is where a quote from one of my favorite technical papers comes in: "We need to salt our analyses with liberal doses of common sense." Even a modestly sized blade might not work for someone who can't pull it through the water fast enough to achieve a desired relative velocity between hull and paddle in the equation. Plus we assumed this relative velocity would remain the same when parametrically looking at the blade area by itself. So, blade force does depend on paddle size. Just be careful about running with this fact in isolation.

The paddle force also varies in direct proportion to the blade's drag coefficient. The drag coefficient is a function of the blade geometry and associated hydrodynamics. Altering the blade's shape to increase its drag coefficient is a research project. So, for the present, let's assume that it too is a constant.

We know from Fig. A.7 that the cosine function has a maximum value of 1, which occurs when its argument equals zero. For the moment, we'll assume that the blade bend angle ϕ is zero. The cosine's argument equals zero when the shaft angle θ is zero. This occurs when the shaft is vertical (and the blade as well, since we've assumed that the bend angle is zero for this straight shaft paddle). If we introduce a bend angle, the difference between the shaft and bend angles must be zero to maximize the cosine. The argument is zero for the bent shaft case when the blade face is vertical.

We know from experience that the shaft angle θ varies with time. This informs our interpretation of the relative velocity since we know from experience that it also depends on time. The fraction, water density, blade area, and drag coefficient are all fixed for a given paddle. The product of the shaft angle cosine and the relative velocity causes the paddle force to vary over time. Since the relative velocity is squared, it is an even function and thus is either zero or a positive number.

So, when is the paddle force maximum? This occurs when the cosine of the angle difference goes to 1, its maximum value (i.e., the blade face is vertical, $\theta = 0$ degrees), when the difference between the hull and horizontal paddle velocities is maximum. This interpretation echoes what rowers are told to do: after loading the blade and starting to pull, focus on hand speed to maximize blade force, especially when the blade face is perpendicular to the water's surface. It turns out the same advice holds for paddlers.

And we read it all in an equation.

Glossary of Terms

1RM   The maximum weight a person can lift once for a given exercise. Also known as a person's single repetition maximum for that exercise. 1RM can vary with training (or lack thereof).

Acceleration   The rate of change of velocity with respect to time. In its most general form, acceleration is a vector, meaning it has both magnitude and direction.

Accelerometer   A device that measures an object's acceleration in one or more directions (axes) when attached to it. A dynamic accelerometer's measurements are tied to the object's reference frame when it and the object move.

Aerobic System   Metabolic pathway that powers muscle contractions via the burning of carbohydrates in the presence of oxygen.

Algebra   The mathematical discipline where computations are performed using symbols, called variables, representing numbers. Rules for manipulating these symbols are grouped together in formulas. After that, one seeks solutions – numbers or variables – that satisfy these formulas. Algebra is a generalization of arithmetic.

Anaerobic System   Metabolic pathway that powers muscle contractions via the breakdown of glucose in the absence of oxygen. Also known as the lactate system.

Antiderivative   An indefinite integral.

Archimedes' Principle   The buoyant force experienced by an object immersed in a fluid equals the weight of the fluid displaced by the object. This is a consequence of Newton's 3rd Law.

Argument   One or more variables that a mathematical function uses to determine the function's value. Also known as an independent variable since the function depends upon it.

Average Force   The propulsive force provided by a paddle averaged over time.

Beam   The maximum width of a hull.

Bend Angle   The angle between a paddle blade's power face and the shaft.

Bent Shaft Paddle   A canoe paddle where the blade's power face and the shaft do not lie in the same plane. Also written as bentshaft paddle.

Blade   The spatulate end(s) of a paddle, wider than the shaft, whose area and drag coefficient enable the paddler to propel a hull.

Blade Angle   The instantaneous angle between a paddle blade's power face and an imaginary line perpendicular to the water's surface. The blade angle is measured in a plane perpendicular to the water's surface and parallel to the hull's keel.

Boundary Layer   The thin layer of fluid adjacent to a hull where the fluid's flow speed varies monotonically from zero at the surface (a consequence of the no-slip condition) to the freestream flow speed. The distance from the hull where the flow speed equals 99% of the freestream flow speed is conventionally defined as the boundary layer's thickness.

Bow   The front end of a hull.

Brace   A static paddle stroke where the blade is used to exert a righting force when a hull heels.

Buoyancy Force   The weight of the fluid displaced by an object, whether floating or submerged.

C1, C2, C4   Solo, tandem, and four-person canoes, respectively

Cadence   The number of strokes performed in a specified amount of time. Customarily, cadence is reported in strokes per minute using a sliding average.

Calculus   The mathematical discipline concerned with change (differential calculus) and accumulation (integral calculus) of functions. Calculus, jointly developed by Isaac Newton and Gottfried Leibniz in the late 17th century, grew from the concept of infinitesimals. Infinitesimals are quantities closer to zero than any number yet do not equal zero. This made it possible to calculate a function's rate of change at a point or compute the area under it.

Canoe   A small, lightweight, and narrow vessel, pointed at one and more often at both ends, paddled by one or more people using single-bladed paddles. Some canoes are decked, either partially or entirely, but more typically are open on top. Outside of North America, these are often referred to as Canadian canoes.

Capsize   See *trout scouting*.

Catch   The first event in the paddle cycle, where the paddle blade is inserted in the water but before the power phase pull.

Center of Buoyancy   The geometric center of the volume of fluid displaced by a hull's submerged portion. This is the point through which the buoyancy force acts.

Center of Gravity   See *center of mass*.

Center of Mass   The location in a distributed object such as a hull or human body through which a force can be applied without causing rotation. If the object's mass is equally distributed throughout its volume, the center of mass equals the object's geometric center.

Center of Rotation   The point about which an object rotates. For stationary hulls with neutral trim and symmetric mass distribution, the center of rotation is the middle of the boat.

Centroid   The arithmetic mean position of all points in a 2-D or 3-D object. Also known as the geometric center.

Chine   A sharp change in the profile of the exterior surface of a hull's cross-section, generally below the waterline. The chine's radius of curvature is notably less than that of the overall hull cross-sectional shape.

Closed-form Solution   An expression comprising mathematical functions and operations that solves a given problem; a solution that can be written down with pencil and paper. Certain limiting cases of the given problem may admit closed-form solutions as well.

Coincident   Occupying substantially the same location, area, or volume; occurring simultaneously.

Conconi Test   A test that measures a person's heart rate at different exercise loads to infer their anaerobic threshold heart rate. The test can be performed on a paddling ergometer over a discrete sequence of increasing speeds or steps. Also known as a step test.

Concrete Water   Neither very shallow nor deep water, with wave drag effects that make it feel like you're paddling in concrete. Concrete water is a qualitative term whose depth depends on the paddler and how tired they are. Some paddlers refer to this as "suck water."

Conservation of Mass   The mass of any closed system must remain constant over time. Consequently, mass can neither be created nor destroyed but can be transformed or rearranged.

Constant   A fixed number, function, or another non-changing mathematical object. Also see the Appendix.

Continuity   The fluid that flows into a control volume over a specified time, minus what flows out, must equal zero unless there is a source there adding fluid or a sink removing it. Continuity is a consequence of the law of mass conservation.

Control Volume   A region with a given volume, typically fixed in space, through which fluid flows.

Coplanar   Lying within the same spatial plane.

Corrective Stroke   A paddling stroke that adjusts or "corrects" the orientation of a hull while underway.

Cross-stream   A direction perpendicular to a collection of streamlines; perpendicular to the direction of flow.

Deep Water   Water that is deeper than half a hull's waterline length.

Deflection Point   The inflection point in a plot of heart rate versus exercise load corresponds to the transition from primarily aerobic metabolism to anaerobic metabolism. Ideally, this corresponds to a blood lactate concentration of 4 millimoles per liter.

Density   A substance's mass per unit of volume.

Derivative   A function's instantaneous rate of change, e.g., the ratio of the change in the function to that of its argument. See also fluxion.

Direction Cosine   Here, the direction cosine is the cosine of the angle between a vector perpendicular to a paddle's power face and the horizontal axis, e.g., the water's surface. The direction cosine is used to compute the instantaneous propulsive force from the force exerted by the blade when paddling.

Displacement Hull   A hull supported exclusively by buoyancy whose hull speed is determined by its waterline length. By contrast, planning hulls are supported by dynamic pressure owing to the speed they travel over the water.

Downstream   In the direction of the current.

Drag, Drag Force   The force that opposes the motion of a hull or other object in a fluid. For paddled hulls, the drag force comprises friction drag, form drag, and wave drag. For paddles it only includes the first two.

Drag Coefficient   A number that quantifies the drag opposing an object's motion in a fluid. Drag coefficients are always associated with a specific hull or paddle. For paddles, the drag coefficient synopsizes the effects of friction drag and form drag; for paddled hulls, it further incorporates the effects of wave drag.

Dragon Boat   A long, narrow watercraft powered by eight to twenty paddlers or more (depending on boat size and class) arranged in pairs. Dragon boats include a drummer who sits in the bow to set the paddlers' cadence, and a steerer.

Duty Cycle, Duty Factor   Here, the fraction of one paddle cycle where the paddler is exerting a padding force, expressed as either a fraction or percentage.

Dynamic Pressure   The pressure arising from fluid in motion. The dynamic pressure equals the fluid's kinetic energy per unit volume.

Dynamic Viscosity   The force required to overcome a fluid's internal molecular friction so it can flow. Also referred to as the fluid's viscosity.

Energy   The ability to perform work. The total energy of a system comprises kinetic or motion energy plus potential energy, e.g., the potential of an object to move. Energy is a conserved quantity, meaning it may be transformed but not destroyed.

Ergometer   A machine for measuring the amount of work performed. An ergometer may display quantities such as pace, distance, force, and time. Also known as an "erg" since an erg is a unit of energy.

Ferrying   Paddling with the keel set at an angle to a river's streamlines, so that the hull's downstream travel is minimized. The current's dynamic pressure causes the hull to translate across the stream in the direction "pointed to" by the bow.

Fluxion   A term introduced by Isaac Newton to describe the instantaneous rate of change of a time-varying quantity or function at a specified point. Nowadays, this is called a time derivative.

Force   An action that can affect an object's motion, causing it to change velocity and thus accelerate. Forces are vector quantities because they have both magnitude and direction. If an object acted upon by a force does not accelerate (e.g., move), then the applied force must be balanced by a reaction force. The reaction force has an equal magnitude and opposite direction to the applied force. This state of balance is called static equilibrium and is a consequence of Newton's 3rd Law.

Force Curve The variation of paddle force versus time in the direction of travel.

Forcing Function An input to a system that produces an output response.

Form Drag The drag force arises from the momentum deficit created by an object's wake. Hulls and paddle blades with a larger cross-sectional area generally have higher form drag. Slender hulls generally have lower form drag; in all cases, the form drag's magnitude is determined by the hull's particular shape.

Freestream Flow The flow upstream of a hull before being deflected and slowed. This also refers to flow far enough to the side of a hull where the water is unaffected by the hull's relative motion.

Free Surface The surface of a fluid having no tangential shear stress in the plane of that surface. Water in a gravitational field forms a free surface when not confined from above.

Freeboard The distance from a hull's waterline to the gunnels, measured at the lowest point of sheer at the top of the gunnels.

Friction Coefficient The ratio of tangential shear stress at the hull's surface to the dynamic pressure. The friction coefficient is inversely proportional to a fractional power of the Reynolds Number.

Friction Drag Drag force that arises due to a fluid's viscosity.

Froude Number The dimensionless ratio between the hull's relative speed and the deep water surface phase speed. The Froude Number is indicative of a hull's wave drag.

Galilean Transformation Geometric transformation between the coordinates of two inertial reference frames.

Geometry The branch of mathematics concerned with spatial properties such as distance, size, and relative position of points, lines, and objects.

Governing Equation A mathematical model that describes how one or more unknown dependent variables change when one or more of its independent variables change.

GPS The Global Positioning System, which provides continuous positioning and timing information anywhere in the world using a constellation of orbiting satellite transmitters and a ground- or air-based receiver. The transmitters broadcast a signal a GPS receiver uses to compute signal transit times from the satellites in view and, thence, the distance to each. The receiver uses this information to infer its latitude, longitude, and altitude.

Grading   Time adjustments used to adjust race results based on age, gender, and hull type.

Gunnel (Gunwale)   The top edge of a hull's side, often incorporating one or more stiffening elements so that the hull maintains its shape.

Heel   Leaning a hull over to one side.

Homogeneous Solution   The solution to a governing equation when the input equals zero.

HRM   Heart Rate Monitor, a device that measures and (typically) displays heart rate data in near real-time. The HRM may also record this data for later analysis.

HRmax   The highest number of heartbeats per minute when exercising at maximal effort. HRmax varies with age and gender.

Hull Speed   For displacement hulls, the speed at which the wavelength of its bow wave equals the hull's waterline length.

Hydrodynamics   The study of fluids in motion.

Hydrostatics   The study of stationary fluids.

Impulse   Here, the time integral of propulsive force over the length of a stroke's power phase. Since force is a vector, impulse is as well. Impulse changes a hull's momentum in the resultant direction, e.g., the direction opposite the propulsive force; this is a consequence of Newton's 3rd Law.

IMU   Inertial Measurement Unit, a device that measures acceleration and rotation.

Independent Variable   See *argument*.

Inertial Reference Frame   A frame of reference that is not accelerating.

Initial Value Problem   An differential equation plus a specified value of the solution at a particular point in time, typically the starting time.

Integral   Mathematical operation that assigns expressions to functions that equal displacement, area, velocity, volume, and the like by combining infinitesimal data.

Integration   The process of computing or deriving an integral.

Interval Workout   Training that involves a series of high-intensity exercise segments interspersed with recovery periods. Interval workouts allow athletes to exercise at higher levels of intensity for longer than they might be able to with a continuous effort.

Inviscid Flow   A fluid that has no viscosity or is idealized as having none; hydrodynamic analysis where viscous effects are not included.

Isotropic   Uniform in all directions; directionless or having no preferred direction.

K1, K2, K4   One-, two-, and four-person kayaks, respectively.

Kayak   A small and narrow decked watercraft, pointed at both ends, paddled by one or more people using double-bladed paddles.

Keel   A hull's bottom-most lengthwise structural element or an imaginary line corresponding to it. The keel is located equidistant from the sides of the hull along its length.

Keel Angle   The angle the keel makes with respect to a river's streamlines.

Kelvin-Froude Waves   The surface wave pattern created by a hull moving in deep water, confined within a 39-degree wedge-shaped area, comprising both transverse and diverging surface waves. Also called Kelvin-Froude waves.

Kinematic Viscosity   The ratio of the dynamic viscosity to the fluid density.

Kinetic Energy   An object's motion energy.

Lactate Threshold   The level of exercise effort where the rate of blood lactate production equals its removal rate. For most exercises, this corresponds to 4 millimoles/liter blood lactate concentration. See also deflection point.

Laminar Boundary Layer   A boundary layer whose streamlines are generally parallel. The fluid there can be idealized as flowing in a stack of parallel laminations. The momentum transfer between adjacent layers is small.

Laminar Sublayer   The region of a turbulent boundary layer very close to the hull and dominated by viscous shear effects. Also called the viscous sublayer.

Lead / Lag   The position in time of an event with respect to a reference time. To lead means the event occurs before the reference, while to lag means the event occurs after it.

Lean   See *heel*.

Lift Force   The component of the force exerted by the fluid flow over an object that is perpendicular to the incoming flow direction.

Limiting Case   The form a mathematical function takes when one or more of its components are extremized. For example, one can set an element to zero when it is small compared to other terms. Or one can set an element to infinity when it is much larger than other terms. For example, in the steady state, such as when a hull is cruising at a constant speed, terms associated with rapid changes in the output can be discarded.

Linear   An expression comprising constants, variables, functions, and mathematical operations where every variable and function is raised to the first power only. The plot of a linear function is a straight line when drawn in a Cartesian coordinate system. Nonlinear functions are sometimes be approximated as linear over specific ranges of their independent variable(s).

Magnitude   The size of a variable or function.

Mass   The measure of an object's resistance to acceleration when a force is applied. This is a consequence of Newton's 2nd Law.

Mass Flow Rate   The mass of a fluid that passes through a specified area or volume per unit of time.

Mean Square Error   The average squared difference between two sets of values, either at discrete points or continuously between functions.

Metacenter   The point where a vertical line running through a rolled hull's center of buoyancy intersects the line that passed vertically through the center of buoyancy at equilibrium. The metacenter characterizes how a hull resists capsize.

Midships   The geometric center of a hull.

Mixed Tandem   A tandem canoe, kayak, surfski, or SUP paddled by two people of different genders.

Moment of Inertia   The measure of an object's resistance to rotational acceleration about a specified axis when a torque is applied. The moment of inertia depends on the object's mass distribution and the axis' location and orientation.

Momentum   The product of an object's mass and velocity. It is a vector quantity, meaning it has both magnitude and direction. In an inertial reference frame, momentum is conserved. If an object is not acted upon by external forces, its momentum does not change.

This is a consequence of Newton's 1st Law. Newton's 2nd Law states that the time rate of momentum change equals the force acting on the object.

Monotonic   Something that is either constant, constantly increasing, or constantly decreasing. The increase or decrease need not be linear.

Net Force   The vector sum of forces acting on an object.

Newton's 1st Law   "Every body continues in its state of rest, or of uniform motion in a straight line, unless it is compelled to change that state by forces impressed upon it."

Newton's 2nd Law   "The change of motion of an object is proportional to the force impressed, and is made in the direction of the straight line in which the force is impressed."

Newton's 3rd Law   "To every action, there is always opposed an equal reaction; or, the mutual actions of two bodies upon each other are always equal, and directed to contrary parts."

No-slip Condition   At a moving hull's surface the water has zero velocity relative to it.

Nondimensional   A quantity or mathematical object that has no physical dimension.

Nonlinear   An expression comprising constants, variables, functions, and mathematical operations where at least one variable or function is raised to any power aside from one. A nonlinear function is a function whose plot is not a straight line when drawn in a Cartesian coordinate system.

Normal   Perpendicular.

Normal Drag Coefficient   A paddle blade's drag coefficient when flow impinges normal to the power face.

Normalize   Dividing a quantity by a specified reference, such as dividing a force by its maximum value so that the resulting normalized force has a maximum of one. A normalized quantity is, in essence, a ratio. When this ratio is small, the numerator is small compared to its denominator, and vice versa when the ratio is large.

Numerical Solution   A solution to a problem derived using numerical approximation, in contrast to symbolic manipulations that provide a closed-form solution.

OC-1, OC-2, OC-6   One-, two-, and six-person outrigger canoes, respectively. Note that OC-1 and OC-2 also refer to solo and tandem open (e.g., non-decked) canoes used in running whitewater. These employ inflatable bags that provide supplemental flotation and prevent swamping.

Optimization   Selection of the best solution with respect to some metric from a set of possible solutions. For example, finding the best fit to a data set by minimizing the mean square error between the data and curves that might fit the data.

Outrigger   A small and narrow watercraft comprising one or more external floatation chambers known as outriggers ("amas"), affixed to one or both sides of the hull via booms. Also called an outrigger canoe.

Overdamped System   A system that responds to external inputs without overshooting its final output value or set point. This is in contrast to an underdamped system, whose output will oscillate about its final value or set point.

Paddle   A handheld object comprising a shaft and one or more spatulate distal ends called blades used to apply force to the water. The shaft has a round or oval cross-section, and is often straight but sometimes curved to facilitate a lower-hand grip with less wrist rotation around the catch. Canoe, outrigger, dragon boat, and SUP paddles have single blades at one end and a grip at the other, while kayak and surfski paddles have two blades.

Paddle Cycle   The four events that correspond to a paddle blade's use. The paddle cycle begins with the catch, followed by the power phase, exit, and recovery.

Peak Force   The maximum amplitude of the force curve.

Phase   The temporal relation between two or more events; the argument of a trigonometric function. See also lead and lag.

Phase Speed   Here, the speed at which any single frequency (or wavelength) component of a surface wave travels.

Phosphate System   System that powers muscle contractions using the energy released when phosphate ions are unbound from stored adenosine triphosphate (ATP). The ATP stores are replenished from stored creatine phosphate (CP), which provides the necessary phosphorous ion to resynthesize it.

Pitch   Movement that raises or lowers the bow via hull rotation about a horizontal axis running midships across the beam. See also Fig. 10.2.

Pivot   The thing or location on which something rotates.

Planing Hull   A vessel supported by dynamic pressure at least some of the time it is underway.

Point Mass   A concept from classical physics, where an object having distributed mass is modeled as being infinitesimally small and located at its center of gravity.

Poling   Propelling a canoe using an 11- to 12-foot aluminum or wooden pole while standing. One plants the pole's lower end at an angle on the bottom of a river or shallow lake, pushes the pole with both hands, then recovers by sliding it up and forward. Using a kayak stroke, a lightweight aluminum pole can propel the canoe over deeper water. Colloquially known as "standing tall in your canoe."

Potential Energy   For mechanical systems, the energy "stored" by an object because of its relative position in a gravity field or from internal elastic stresses.

Power   The amount of energy transferred or converted in a system per unit of time.

Power Curve   The plot of applied power versus average velocity for moving a range of weights in a strength training exercise. The curve's inflection point corresponds to the average velocity, which generates maximum power for that exercise.

Power Face   The side of a paddle blade facing opposite the direction of motion.

Power Phase   The portion of the paddle cycle over which the paddle is pulled to provide propulsion.

Pressure   The compressive stress within a fluid. Pressure is isotropic and scalar. It exerts a normal force distributed over solid surfaces in contact with the fluid.

Pro Boat   Tandem canoe hull built to United States Canoe Association (USCA) specifications, 18.5 feet long and 27 inches wide at a waterline 3 inches above the bottom of the keel.

Propulsive Force   Here, the projection of the force exerted by a paddle blade in the direction of travel.

Recovery   The portion of the paddle cycle when the paddle is out of the water and moving forward in preparation for the ensuing catch.

Recovery Fraction   The percentage of the paddle cycle comprising the exit, recovery, and catch.

Reference Frame   A coordinate system whose origin, orientation, and scale are specified by a reference location.

Refraction   Here, the redirection of a water wave as it travels from one depth to another.

Relative Velocity   Here, the difference between the velocity component of the paddle blade's power face in the direction of travel versus the hull velocity.

Restoring Force   A force that restores an object to its equilibrium position.

Reynolds Number   The nondimensional ratio between fluid inertial and viscous forces acting on a body. The ratio is written as the product of the flow speed times a characteristic length scale, such as the distance from the bow, divided by the fluid's kinematic viscosity.

RF   Radio Frequency, often used as shorthand for wireless radio devices and systems.

Ricatti Equation   A family of first-order nonlinear ordinary differential equations, quadratic in the unknown function, that do not generally admit a closed-form solution. Ricatti equations are most often solved numerically. Every linear second-order homogeneous ordinary differential equation with variable coefficients can be reduced to a Riccati equation by a change of variables.

Righting Arm   The horizontal distance between the buoyancy force vector and a vertical line passing through the hull's center of gravity.

Rocker   The upward curve of a hull's keel line toward the bow and/or stern when viewed from the side. Rocker can be symmetric front-to-back or asymmetric.

Roll   Movement that rotates the hull about a horizontal axis running from bow to stern. See also Fig. 10.2.

RPE   Rating of Perceived Effort, a qualitative measure of exercise intensity.

Rudder   A control surface used to steer a vessel that induces yaw owing to forces acting on it when rotated.

SAR   Speed Above Replacement, or the speed advantage/disadvantage attributable to one paddler in a multi-paddler hull.

Scalar   A quantity unaffected by changes in reference frame because it lacks directionality.

Setting   Ferrying, but with the stern pointing upstream.

Shaft   The tubular portion of a paddle having a circular or oval cross-section.

Shaft Angle   The angle between the shaft and a line perpendicular to the water's surface in a plane parallel to the keel line. If the shaft is not straight, the shaft angle may be referenced to the portion adjacent to the shoulder.

Shallow Water   Water shallow enough where the surface wave speed is independent of wavelength.

Sheer The curvature of the gunnels when viewed from the side.

Shoulder The portion of a paddle where the blade transitions into the shaft.

Simplifying Assumption An assumption that makes a mathematical problem tractable, often by eliminating terms much smaller than others.

Simultaneous Equations Two or more algebraic equations that share at least one independent variable.

Single Repetition Maximum See 1RM.

Sink An element within a control volume that removes fluid.

Solo Here, a hull paddled by a single individual.

Source An element within a control volume that adds fluid.

Spanwise The direction across a river's width, or across a flow.

Specific Impulse Impulse normalized by the paddler's mass.

Speed The magnitude of velocity.

SPM Strokes per minute. Also see cadence.

Stem One end of a hull, either the bow or stern.

Stern The back end of a hull.

Streamline The set of curves whose tangents comprise the velocity field of a fluid flow.

Stroke Duration The time between consecutive occurrences of the same element of the paddle cycle, such as the time between two successive catches or two consecutive exits. Stroke duration equals the inverse of the instantaneous cadence.

Sublayer Bursting An upwelling in the laminar sublayer which acts like a localized flow separation, feeding the turbulence in a turbulent boundary layer.

SUP Stand-Up Paddleboard, a surfboard-like watercraft where one or more paddlers stand on the board's deck and propel it using a single-bladed paddle.

Superposition The principle where the net response caused by two or more inputs is the sum of the responses caused by each input separately. This property holds for all linear systems, including small-amplitude water waves.

Tandem   Here, a watercraft paddled by two paddlers.

Taylor Series   A means for representing a function via an infinite sum of terms comprising the function's derivatives at a specified point.

Tollmien-Schlichting Waves   An unstable pressure wave that propagates in the stream-wise direction within a laminar boundary layer, driving the boundary layer's transition to turbulence.

Total Drag Force   See drag force.

Training Effect   The result of training at progressively higher training loads and/or volumes. When the training load is decreased, one's fitness level grows to surpass their initial fitness. A decreasing fitness level with an increased training load may indicate overtraining.

Transition   The portion of a boundary layer between the laminar and turbulent regions where the laminar boundary layer's parallel streamlines become disordered.

Transverse   Across.

Trigonometry   The branch of mathematics concerned with the relationships between the dimensions and angles of triangles.

Trim   The equilibrium pitch angle of a hull. Neutral trim means the hull's pitch is zero degrees, referenced to the water's surface. Bow-down trim means that the hull is pitched downwards toward the bow at equilibrium. Bow-light trim means that the hull is pitched upwards toward the bow at equilibrium. Static trim is the pitch set when the hull is stationary; dynamic trim is when the hull is underway. Shallow water effects can impact dynamic trim, pulling the stern downwards while raising the bow.

Trout Scouting   Falling out of a boat, which enables the paddler to search for fish. Sometimes referred to as a wet exit.

Turbulent Boundary Layer   A boundary layer whose streamlines are disordered, with significant transfer of fluid momentum over large portions of its thickness and width.

Upstream   Opposite the direction of the current.

Variable   A symbol that represents a quantity in a mathematical expression. Unlike a constant, a variable does not have a fixed value. Also see the Appendix.

VBT Velocity Based Training, a strength and power training method based upon the speed at which a weight is moved. VBT allows an athlete to functionally strength train cyclic or acyclic movements irrespective of variations in 1RM.

Vector A mathematical object having both a magnitude and a direction.

Vector Component The projection of a vector on one of a reference frame's coordinate axes. A vector can be expressed as the sum of its components in a multidimensional reference frame.

Velocity The rate of change of position with respect to time. In its most general form, velocity is a vector, meaning it has both magnitude and direction. By contrast, speed has magnitude but no direction.

Velocity Deficit See *wake deficit*.

Viscosity In fluid flow, another term for kinematic viscosity.

VO2max The maximum rate of oxygen consumption one can attain during physical exercise. Also written as VO2 max.

Vorticity The spinning motion of a fluid around some point.

vVO2max Velocity (speed) when paddling at VO2max.

Wake The region of recirculating flow behind a moving hull.

Wake Deficit The difference between the flow's freestream speed and the average speed in a wake.

Waterline The horizontal line where a ship's hull meets the water's surface.

Wave Drag The drag force that arises because of energy loss due to wave creation by a moving hull.

Wavelength The spatial period of a wave; the distance over which the wave's shape repeats.

Weight The force acting on an object in a gravitational field due to its mass.

Wetted Area The surface area of a hull below its waterline.

Work The energy transferred to or from an object through the application of force over a distance.

Yaw   Movement that pushes the bow side-to-side about a vertical axis located around midships passing through the keel. See also Fig. 10.2.

List of Symbols

a	acceleration
A	area
$\mathbf{B}$	center of buoyancy
c	local current magnitude
$\mathbf{C}$	current vector
C	current magnitude
C_d	blade drag coefficient (non-dimensional)
C_D	hull drag coefficient (dimensional)
c_p	phase speed
cos	cosine function
d	distance
dA	differential area
D	depth; average depth; duty cycle
e	mean square error
E	energy
f	function – *see also* Appendix
F	force
F_D	drag force
F_p	propulsive force
F_r	Froude Number
g	acceleration due to gravity
$\mathbf{G}$	center of gravity; center of mass
h	depth; local or instantaneous depth; time increment
J	impulse
k	Runge-Kutta approximation elements
L	length
m	mass
M	river half width; mass (combined mass)
$\mathbf{M}$	metacenter
N	number of elements in a summation
p	momentum; pressure
P	power; momentum
R	radius

$\mathbf{R}$	resultant vector
R_e	Reynolds Number
s_i	velocity vector component
$\mathbf{S}$	velocity vector
S	velocity vector magnitude
sin	sine function
t	time
T	period; time interval
tanh	hyperbolic tangent function
u, U	speed
v, V	speed
w	speed
W	weight; work
x	coordinate direction
y	coordinate direction
z	coordinate direction
δ	laminar sublayer thickness
Δ	change in a quantity or variable
θ	angle; shaft angle
λ	wavelength
ν	fluid kinematic viscosity
π	pi, the ratio of a circle's circumference to its diameter
ρ	density
ϕ	angle; offset angle; phase shift

Index

1RM, 179–180, 182

A

acceleration, 4–6, 41, 52, 62, 73–74, 90, 125, 133–134, 186, 225–226, 231
ADP, 170, 174
aerobic system, 168–172, 174–175, 191, 193
anaerobic system, 172–173, 175
anaerobic threshold, 195, 197, 199–200
Archimedes' Principle, 9–10, 53, 113
argument, 30, 39, 67, 107, 149, 151, 218
ATP, 169–170, 172–174, 183–185

B

bend angle, 61, 67–68, 70, 76, 82, 232
Bernoulli analysis, 23–24, 26
Bernoulli's Principle, 24, 41
Billat 30-30, 203, 205–207
Billat, Veronique, xvi, 203
blade angle, 61–62, 68, 82
blood lactate, 172–173, 175, 188–189, 193–195, 199–200
Bolt, Richard, xiii, 93
Bompa, Tudor, 179, 187
boundary layer, xi, 18–22, 28–29
  laminar boundary layer, 19, 23
  transitional boundary layer, 18–19
  turbulent boundary layer, 20, 28
buoyancy force, 5, 52–54, 56–59

C

cadence, xiii, 64, 71–72, 76, 79–83, 137–138, 140, 182
Caplan, Nicholas, 63, 70, 85, 132
capsize, 51, 58–59
catch, 44, 63–65, 68, 70, 75, 81–83. *See also* paddle cycle
center of buoyancy, 53, 55–56, 58–59
center of gravity, 54, 56–59, 117, 121
center of mass, 52–56, 58–59, 117, 122
center of rotation, 103, 121
characteristic area, 13–14

Clark, Kenneth, 72, 85
closed-form solution, 125, 134
Conconi, Francesco, 194
Conconi test, 194–195, 199
concrete water, 39, 152
conservation of energy, 36, 38, 42, 49
conservation of mass, 24, 30. *See also* continuity
conservation of momentum, 44, 48, 64, 73, 138
continuity, 30–31. *See also* conservation of mass
control volume, 29–31

D

deep water, 34–35, 42, 126, 144, 146, 149, 151, 157
deflection point, xv, 188, 194–195, 198–199
density, 9, 13–14, 27, 30, 49, 51, 62, 66, 78, 82
derivative, 62, 74, 90, 125, 133–134, 225
dimensional analysis, 12–13
direction cosine, 63, 67–69
displacement hull, 36
drag, xi, xii, xiv, 3, 7–10, 12–14, 16–17, 21, 23, 25–29, 31, 60–62, 73, 83–85, 90–92, 115, 123, 128, 130, 134, 136, 140, 155, 189, 194, 198–199
  form drag, xi, xii, 4, 17, 21, 23, 27–29, 31, 60, 70, 115, 120, 122, 144
  friction drag, xi, 16–23, 28–29, 70, 115, 122
  total drag, 21, 29, 36, 83, 130
  wave drag, xi, xiv, 17, 21, 33, 36, 39–41, 60, 144–145, 148
drag coefficient, xiv, 14–15, 62–63, 66, 78, 82, 84–85, 91, 124–126, 130–131, 134–135, 212, 232–233
Drag Equation, 12, 14, 115. *See also* dimensional analysis
drag reduction, 28
  air injection, 28
  LEBU, 28
  polymer injection, 16, 28
  riblets, 28
duty factor, 77, 79–83, 135–138, 140–141
dynamic pressure, 3, 14, 31, 115, 117, 122–123

About the Author

As an engineer I find paddling not only fun, but technically fascinating. Spending hours at a time on the ergometer in Winter, and on the water when it's not frozen here in New England, has given me time to ponder why canoes do what they do.

I'm an avid paddler, having first explored the sport as a wilderness canoe tripper. I'm a dedicated canoe racing middle-of-the-packer, and have been known to pole canoes, run spring whitewater, and compete in canoe orienteering. I was an instructor and organizer at the Maine Canoe Symposium from 1997 to 2018. When I'm not paddling or writing, I work as an engineering and intellectual property consultant. I'm an inventor and patent holder of technologies for sensing and sports performance monitoring, with ten issued patents and numerous conference presentations and journal publications in a variety of fields.

I hold a BSE *cum laude* in mechanical and aerospace engineering from Princeton University, and an MS and PhD in mechanical engineering from the Massachusetts Institute of Technology with a minor in applied mathematics. I've worked, studied, and conducted research in acoustics, fluid mechanics, embedded systems, wired and wireless sensing, biophotonics, and control systems.

And in case you're wondering, I wrote this book because it's fun. All of it.

Join us

The Science of Paddling blog:
www.thescienceofpaddling.net

Like or Follow us on Facebook:
facebook.com/thescienceofpaddling